Springer Geography

For further volumes:
http://www.springer.com/series/10180

The Springer Geography series seeks to publish a broad portfolio of scientific books, aiming at researchers, students, and everyone interested in geographical research. The series includes peer-reviewed monographs, edited volumes, textbooks, and conference proceedings. It covers the entire research area of geography including, but not limited to, Economic Geography, Physical Geography, Quantitative Geography, and Regional/Urban Planning.

Cerasella Crăciun · Maria Bostenaru Dan
Editors

Planning and Designing Sustainable and Resilient Landscapes

Springer

Editors
Cerasella Crăciun
Urban and Landscape Design Department
"Ion Mincu" University of Architecture and Urbanism
Bucharest
Romania

Maria Bostenaru Dan
Urban and Landscape Design Department and Centre for Architectural and Urban Studies

and

"Ion Mincu" University of Architecture and Urbanism
Bucharest
Romania

ISSN 2194-315X ISSN 2194-3168 (electronic)
ISBN 978-94-017-7931-9 ISBN 978-94-017-8536-5 (eBook)
DOI 10.1007/978-94-017-8536-5
Springer Dordrecht Heidelberg New York London

Printed on acid-free paper

Springer is part of Springer Science+Business Media (www.springer.com)

On the occasion of 10 years of Landscape Design and Planning Department teaching at the

Romanian "Ion Mincu" University of Architecture and Urban Planning in Bucharest,

dedicated to all specialists and students in the field

Preface

Rapid industrialization, the accelerated rhythm of everyday life and the decrease of collective life quality have intensified more as the society moves towards faster informational and technological systems putting the landscape is in a delicate position. Therefore, the actual landscape comprises all existent past or possible images and it is under the imperative of evolution law, from landscape of intervention at micro and urban scale of to the evolving landscape at macro-scale, maintained by human societies from economic, political and socio-cultural reasons.

This book aims to strengthen the knowledge base dealing with landscape in all its components: the green space and the built landscape as material part, but also the immaterial landscape of conserved traditions. We included both scales, the territory and the detail, landscape urbanism and landscape architecture, planning and design, with a slight overrepresentation of the macro-scale.

We differentiate ourselves from numerous books dealing with cultural landscape and have chosen instead sustainable and resilient landscapes as a response to environmental challenges posed by the transformation of living ways in the twenty-first century. Cultural landscapes are past and present shaped by man and nature, while with sustainable we aim to a look to the future, emphasised also by the focus on the 'project' ('Entwurf' in German): planning and design. Resilience instead aims to include the cultural dimension. Resilience planning aims at a sustainable response to environmental challenges. Resilient communities shall be built, and here we find the man–nature interaction of the cultural landscape.

The book is containing research articles which contain a critical view to landscape architecture and landscape urbanism practice and can as well help practitioners to question their methods in today's world where the planner is the moderator between numerous disciplines dealing from the historical background to the design of the future and the communication to users. In a next volume, we will highlight also our approach to education and teaching by research.

The book brings 17 chapters dealing with first with landscape archetypes, defining cultural landscape as identity of the community shaping it as first two cross-topic parts. The next parts are dealing landscape in a transition from the natural landscape to the landscape of buildings and open spaces defined by them: ecological landscape, transition landscape of urban peripheries, urban gardens as detailed view, urban landscape and public space. The aims of the editors were to select contributions describing various 'scapes' and do not restrict the book on one specific scale. The idea behind this approach was, that a great variety of ways describe the presence of landscape in our lives and complexity of approaches are needed to understand the present state and future changes in landscape. The authors are experts with various backgrounds in the field of architecture, urbanism, geography, ecology, history of art, communication science. This multidisciplinary approach allowed the coverage of landscape-related issues from various aspects.

For the part title, we first gave a described category of 'scape', then we selected some keywords common to the chapters in the part, and finally a 'finality' related to envisaged conservation or intervention. The parts go from first the abstract, then the nature in itself until the mineral city.

The first part of the book is the introduction to the topic seen from the point of view of more disciplines. The last part of the book draws the conclusions by establishing the connections between the chapters. There are numerous links between them, while the division into parts reveals grouping. The parts contain usually two chapters, at most three. The part II of the book is focused on the philosophic approach and describes symbols in understanding and conceiving landscape. It contains two chapters, making the transition to the parts to follow and approaches expressions of immaterial landscape. The part III contains three chapters on the topic of (re)discovering memory and identity in landscape. Two chapters are dealing with interpreting historical and contemporary photography, respectively. The part IV, which covers the technical side of the book, deals with the transformation of the natural landscape, the forest, through human activity, not necessarily directed to the creation of the urban settlements with which the following parts deal. The part V deals with the so-called 'Zwischenstadt' (in German "city in between")—the area between urban and rural. The city is separated from the 'land' today, and the following parts deal with how to gain access to items from this wide landscape in the city again. The chapters till now dealt with rural, agricultural, forest landscape, and from here on the built substance intervenes. This 'scape' of the urban periphery needs new values to be associated with it to gain a (new) landscape character. The part VI goes in detail as compared to the macro-scale dealt with so far. It depicts the relationship between the house and the garden in three chapters. Architects and landscape architects cooperated in history in associating plants with buildings, and today we have innovative ways of even incorporating the plant in the building with green walls. These are only a way of creating green in crowded urban areas, other ways, of (pocket) parks are explored. The part VII is again at the larger scale, this time of the city. It focuses on 'circulation' which has been one of the zones of the city stipulated by the Athens Charter and seen equal to more surface described one as those associated to other functions. One of the chapters deals with the routes of pedestrian circulation, which might be informal and different from the street network and reflect the perception of the image of the city in landmarks and connected elements as described initially in the decoding of landscape in the part II. The second chapter deals with the street network for motorised or non-motorised circulation and how this can get garden elements. Both deal with mobility, as mean for perceiving and defining landscape in investigation of the user or with intervention on mobility policies to make them environmentally more sustainable. The closing group of three chapters deals with the mineral part of landscape: the open space ('Freiräume' in German "free spaces"). It features an interview which brings us back to the philosophic view introduced at the beginning, an overview of the application of approaches from seminal books in urban regeneration and finally communicating messages through media landscape, when the facade becomes the public space as before the facade became the garden: the vertical public space.

Landscape architecture is a new profession which emerged from planting design when the concept of 'cultural landscape' was defined. One chapter of the book is the result of a project showing the beginnings of the profession in Austria—it was in the interwar period and it featured strong participation of women, a rare fact at that time. But it was the Hungarian Mihály Mőcsényi, Sir Geoffrey Jellicoe Awardee, the highest distinction of the International Federation of Landscape Architects (IFLA), in 2012 who defined the contribution of architecture to making landscape architecture what it is, in contrast with the view, which we still give account of in our book, landscape of pure nature. To cite the motivation of the Sir Geoffrey Jellicoe award, in 1968 he [Mihály Mőcsényi] officially defined landscape as 'cultural product' and defined landscape as 'humanized nature'. This established the basis of landscape architecture and planning in Hungary as an integrative and ecological approach—a revolutionary change from the former the purely geographical landscape approach. For this reason he studied also architecture in addition to his basic studies.

Our book features mostly contributions of architects from the "Ion Mincu" University of Architecture and Urbanism teaching landscape, among other disciplines related to landscape, and is another contribution from Eastern Europe. It appears on the occasion of the 10th anniversary of landscape urbanism teaching at the "Ion Mincu" University of Architecture and Urbanism. Those graduating from the "Ion Mincu" University of Architecture and Urbanism in the study directions led by Cerasella Crăciun are not landscape architects or landscape engineers but landscape urbanists ('urbanist peisagist' in Romanian). Because today a new term is emerging, that of 'landscape urbanism' a term by Peter Connolly from 1994, a theory of urban planning aiming to organise the cities' landscape instead of designing its buildings. We included this view in our book by looking, increasingly towards the end, more to the cities' landscape than to the 'humanised nature' of the rural, agricultural and forest landscape. We look forward for further collaboration, through collaborations initiated by Maria Bostenaru Dan, with the university where the first view on landscape architecture, today spread in the whole world, initiated.

Bucharest, October 2013

Cerasella Crăciun
Maria Bostenaru Dan

Acknowledgments

The idea of this book grew out of now 10 years of landscape architecture and urbanism formal teaching and research at the “Ion Mincu” University of Architecture and Urbanism, Romania. However, landscape has been an issue in Romanian urban planning since half a century, Radu Laurian starting to deal with this in 1962.

We further invited international experts who are linked to it, whom we know through common research stays at the Canadian Centre for Architecture in Montreal, Canada (Stephen Monteiro), through workshops of the Romanian Ministry of Culture in the field of cultural tourism to resort architecture (Iris Meder) or through the teaching in the doctoral programme of the university about participative approaches to landscape (Stephanie Brandt). We further enjoyed the support of the international community of researchers in the review of the chapters, as further acknowledged. We are grateful for the experience at the Canadian Centre for Architecture in dealing with historic photography, which, along with a focus of the doctoral school on the study of image through a consortium with the University of Bucharest including its dedicated research centre reflected among others in courses on photography critique, resulted in an important part of the book dedicated to the photography of landscape.

This first volume focuses on how to deal with landscape between conservation and intervention, depending on what the contemporary state is compared to the one historically planned. Sometimes this conservation can be only virtual, by means of history records, photography or digital media. We envisage to publish further volumes on landscape design, dealing maybe with strategy and visualisation for research.

The volume creates a dialogue between the micro-landscape in the city, the architecture linked by its means, and the macro-landscape at the urban periphery or even more distant than this, the agricultural and rural landscape. Most sites are located in Eastern Europe, in Romania, but in some cases the international spread of the studied object has been considered.

We would first like to thank the authors who participated in this editorial project. Without their high quality contributions and patience during the long publication process, the book would have not been possible.

We would like to acknowledge the following reviewers:

Deirdre Gilfedder (France), Andrea Haase (Germany), Alenka Poplin (Germany), Bas Pedroli (The Netherlands), Thomas Panagopoulos (Portugal), Bob Jarvis (UK), Chris Emrich (USA), Mircea Enache (USA), Thaisa Way (USA), Valentin Anton (Romania), Iuliana Armaş (Romania), Adrian Majuru (Romania), Şerban Raicu (Romania), Ioana Siminea (Romania), Florin Vartolomei (Romania), and an anonymous reviewer.

We also thank to the publishing assistance of the publishing editor Dr. Robert Doe and of the editorial assistants Robert van Gameren and Naomi Portnoy, for the very kind guidance and the patience during the long production time of this volume.

Bucharest, October 2013

Contents

Contributors

Iuliana Armaş Faculty of Geography, Department of Geomorphology-Pedology-Geomatics, University of Bucharest, Bulevardul Nicolae Balcescu no. 1, Sector 1, 010041 Bucharest, Romania, e-mail: iulia_armas@geo.unibuc.ro

Maria Bostenaru Dan Faculty of Urban Planning, Urban and Landscape Design Department and Centre for Architectural and Urban Studies, "Ion Mincu" University of Architecture and Urbanism, Academiei 18–20, 010014 Bucharest, Romania, e-mail: maria.bostenaru@iaim.ro

Stephanie Brandt Stephanie Brandt Studio, Marzellenstraße 43b, c/o Aclewe, 50668 Köln, Germany, e-mail: mail@stephaniebrandt.net

Livioara Braşoveanu Faculty of Geography, Department of Geomorphology-Pedology-Geomatics, University of Bucharest, Bulevardul Nicolae Balcescu no. 1, Sector 1, 010041 Bucharest, Romania, e-mail: brasoveanulivioara@yahoo.com

Oana Andreea Căplescu Faculty of Architecture, Department of Introduction to Architectural Design, "Ion Mincu" University of Architecture and Urbanism, Academiei 18–20, 010014 Bucharest, Romania, e-mail: andreea.caplescu@uauim.ro

Cristina Constantinescu Faculty of Urban Planning, Department of Urban and Landscape Design, "Ion Mincu" University of Architecture and Urbanism, Academiei 18–20, 010014 Bucharest, Romania, e-mail: constcristina@yahoo.com

Cerasella Crăciun Faculty of Urban Planning, Urban and Landscape Design Department, "Landscape and Territory" Masteral Program, "Ion Mincu" University of Architecture and Urbanism, Academiei 18–20, 010014 Bucharest, Romania, e-mail: cerasellacraciun@gmail.com; www.cerasellacraciun.ro

Alex Dill Department of Architecture, Institute for Architectural Design, Art and Theory, Karlsruhe Institute of Technology, Englerstr. 7, 76131 Karlsruhe, Germany, e-mail: alex.dill@kit.edu

Cristina Enache Faculty of Urban Planning, Department of Urban and Landscape Design, "Ion Mincu" University of Architecture and Urbanism, Academiei 18–20, 010014 Bucharest, Romania, e-mail: cristinaienache@yahoo.com

Mihaela Hărmănescu Faculty of Urban Planning, Department of Urban and Landscape Design, "Ion Mincu" University of Architecture and Urbanism, Academiei 18–20, 010014 Bucharest, Romania, e-mail: mihaela.harmanescu@gmail.com

Ulrike Krippner Department of Landscape, Spatial and Infrastructure Sciences, Institute of Landscape Architecture, University of Natural Resources and Life Sciences Vienna, Peter-Jordan-Straße 82, 1190 Vienna, Austria, e-mail: ulrike.krippner@boku.ac.at

Iris Meder Department of Landscape, Spatial and Infrastructure Sciences, Institute of Landscape Architecture, University of Natural Resources and Life Sciences Vienna, Peter-Jordan-Straße 82, 1190 Vienna, Austria, e-mail: iris.meder@boku.ac.at

Stephen Monteiro The American University of Paris, 20 rue Rouvet, 75019 Paris, France, e-mail: stevemonteiro@gmail.com

Mihaela Hermina Negulescu Faculty of Urban Planning, Urban and Regional Planning Department, "Ion Mincu" University of Architecture and Urbanism, Academiei 18–20, 010014 Bucharest, Romania, e-mail: mihaela_negulescu@yahoo.com

Gabriela Osaci-Costache Faculty of Geography, Department of Geomorphology-Pedology-Geomatics, University of Bucharest, Bulevardul Nicolae Balcescu no. 1, Sector 1, 010041 Bucharest, Romania, e-mail: gabrielaosaci68@yahoo.com; gabrielaosaci@libero.it

Alexandru-Ionuţ Petrişor Faculty of Urban Planning, Department of Urban and Landscape Design, "Ion Mincu" University of Architecture and Urbanism, Academiei 18–20, Sector 1, 010014 Bucharest, Romania; National Institute for Research and Development in Constructions, Urbanism and Sustainable Spatial Development, URBAN-INCERC, Soseaua Pantelimon nr. 266, Sector 2, 021652 Bucharest, Romania, e-mail: alexandru_petrisor@yahoo.com

Andreea Popa Faculty of Urban Planning, Department of Urban and Landscape Design, "Ion Mincu" University of Architecture and Urbanism, Academiei 18–20, 010014 Bucharest, Romania, e-mail: apopa.uauim@gmail.com

Angelica Stan Faculty of Urban Planning, Department of Urban and Landscape Design, "Ion Mincu" University of Architecture and Urbanism, Academiei 18–20, 010014 Bucharest, Romania, e-mail: angelicastan@yahoo.com

Part I
Introduction: Research/ Method/Transdisciplinarity—*Metabolism*

Pluridisciplinarity, Interdisciplinarity and Transdisciplinarity: Methods of Researching the Metabolism of the Urban Landscape

Cerasella Crăciun

Abstract The complexity of the world outlook nowadays annihilates the pyramid of the classical outlook of articulating the subjects which consider physics domain as a basis, creating a real disciplinary subject "bing-bang". The field of every subject is getting more and more restricted, which makes today the dialogue among them more and more difficult, if not impossible. The need for bridges among various fields and subjects has materialized itself in the appearance at the end of the twentieth century, of the pluridisciplinarity and interdisciplinarity concepts, as a complex phenomenon, generator of innovation in the field of the urban landscape metabolism as well by fragmentation of subjects, recombining them, diffusion of concepts and resorting to methods from other fields of activity. At the beginning of the new millennium, however, the interdisciplinary approach was rediscovered and used as an answer to the unprecedented challenge launched by the world we live in a "manifesto" of what was among and beyond the subjects, in a new approach, in order to express the need to get over the frontiers among subjects, expressing the need for unity both within the urban field and the landscape one. The methodology of transdisciplinary research is determined by Levels of Reality, The Logic of Included Tertiary and Complexity, which are at the same time the three postulates of modern science, unchanged from Galileo to our day, in spite of the appearance of the infinite diversity of methods, theories or patterns that have covered the history of various scientific fields of activity. Disciplinarity, pluridisciplinarity, interdisciplinarity and transdisciplinarity are complementary, representing "the four arrows of the one and the same bow: that of knowledge", making reference to the idea of scientific integration, to the "great book of nature", in which every scientific subject or art is responsible for a segment of the whole. Transdisciplinarity is

C. Crăciun (✉)
Urban and Landscape Design Department, "Ion Mincu" University of Architecture and Urbanism, Academiei 18–20, 010014 Bucharest, Romania
e-mail: cerasellacraciun@gmail.com

C. Crăciun and M. Bostenaru Dan (eds.), *Planning and Designing Sustainable and Resilient Landscapes*, Springer Geography, DOI: 10.1007/978-94-017-8536-5_1,

complementary to the disciplinary approach out of the confrontation among fields and subjects bringing about new results and new bridges among them, offering a new view on nature and reality.

Keywords Pluridisciplinarity • Interdisciplinarity • Transdisciplinarity • Philo sophy • Science • Research • Method

In the classical view of the world, the process of articulating disciplines was considered to have a pyramidal shape, the basis of the pyramid being represented by physics. However, the complexity of current times annihilates this pyramid, bringing about a genuine disciplinary big bang.[1] Nowadays, the field of each discipline is becoming increasingly limited, which makes the dialogue between disciplines increasingly more difficult if not impossible. The need for *bridges* between different disciplines resulted in the appearance towards the end of the twentieth century of the concepts of ***pluridisciplinarity*** and ***transdisciplinarity***, two concepts that will be studied and applied in this research.

Pluridisciplinarity means ***researching an object from the perspective of one and the same discipline through several disciplines at once***.[2] Thus, the object of research will be enriched as a result of it being related to other disciplines, and therefore, pluridisciplinary research brings an extra something to that discipline. **Interdisciplinarity** involves ***the transfer of methods from one discipline to another*** and is a complex phenomenon generating innovation in the field of the metabolism of the urban landscape by means of ***the fragmentation of disciplines, their recombination, the diffusion of concepts and the process of borrowing methods*** from other disciplines.

The term **transdisciplinarity** emerged three decades ago in the works of different researchers such as *Jean Piaget, Edgar Morin, Eric Jantsch* to **express the need to overcome the borders between disciplines** and referred to (…what was at the same time in between, within and beyond disciplines. Its purpose is to understand this world, one of its imperatives being the unity of knowledge—*translation of original quoted below*).[3] The idea is important because it expresses the need for unity in the field of urbanism, both within the discipline of urbanism and in relation to other contact disciplines.

At the beginning of the new millennium, the transdisciplinary approach is rediscovered and used in response to the unprecedented challenges launched in the world we

[1] **Basarab Nicolescu**, *Transdisciplinaritatea. Manifest.* (Transdisciplinarity. A Manifesto.), *Polirom* Publishing House, *Idei Contemporane* Collection, *Plural* Series, Iasi, 1999, 42.

[2] ***Example***: A Renaissance painting can be studied from the perspective of the history of art, of the history of architecture and urbanism, landscape and nature, from the point of view of physics, of chemistry, of the history of religions, of the history of Europe and of geometry; Marxism can be studied from the point of view of philosophy, economy, psychoanalysis, physics or literature.

[3] Op. Cit., 53. "…*ceea ce se afla în acelaşi timp şi între discipline, şi înlăuntrul diverselor discipline, şi dincolo de orice disciplină. Finalitatea sa este înţelegerea lumii prezente, unul din imperativele sale fiind unitatea cunoaşterii*".

live in as in a “global village”. The Romanian Basarab Nicolescu[4] launches in Paris “*Manifestul transdisciplinarității*” (*The Manifesto of Transdisciplinarity*), which gives a credible answer to the major problems of the contemporary world by means of the generous **unification of Science, Culture and the Sacred**. The transdisciplinary research initiated by the Romanian physicist recognizes the configuration of various levels of existence which represent more than just a solid theoretical discourse on the world, on culture and on the human spirit and proposes *an action programme for the future*.

Transdisciplinarity[5] is a new way of initiation; it **integrates the foundations of the old esoteric traditions and of contemporary science, renewing their language; this is a visionary and operational way** that addresses the most open and awakened consciousnesses and that draws strict lines of action. By establishing bridges between exact sciences and humanities, between science and tradition, between the scientific way of thinking and the symbolic way of thinking and between knowledge and being, transdisciplinarity tends towards the unity of knowledge, going through the mandatory step of self-knowledge.

In the same direction, other Romanian thinkers have made major contributions: (Science without conscience is only the ruin of soul. I see in Transdisciplinarity the Great Breakthrough of the twenty-first century—*translation of original quoted below*) said **Michel Camus**.[6]

---

[4] Basarab Nicolescu is a physicist and a researcher at CNRS. He obtained his doctorate in physics at Université Pierre et Marie Curie, Paris. He specialized in elementary particle theory, he is the author of many scientific articles published in international journals, and he has numerous contributions to joint scientific works. As a result of his interest in the relationships between art, science and tradition, he published works and articles on the role of science in contemporary culture. Founding President of the *Centre International de Recherche et Etudes Transdisciplinaires* (CIRET); cofounder of the *Reflection Group on Transdisciplinarity* affiliated to UNESCO (1992). In 1986, he received the Silver Medal of the French Academy for his work “*Nous, la particule et le monde*”. For the book “*Science, meaning and evolution. The cosmology of Jakob Boehme*” he was awarded in 1992 the Benjamin Franklin Award for Best History Book and the Writers’ Union Prize in Romania in 1993. Books published in Romanian: “*Noi, particula şi lumea*”; “*Transdisciplinaritatea. Manifest; Teoreme poetice*”; “*Ştiinţa, sensul şi evoluţia. Eseu asupra lui Jakob Böhme*”.

[5] There are correspondences between transdisciplinarity and other trends: at the level of experimental metaphysics and transdisciplinary isomorphism, between the outside world and the inner world with the ***Great Game*** (*Le Grand Jeu*)—a lesser known literary movement inspired by surrealism and having an esoteric and gnostic tint—a reaction against basic materialism and against the positivism of exact sciences (representatives: *Roger-Gilbert Lecomte* and *Rene Daumal);* or with the German ***Naturphilosophie*** movement (centred around the journal *Atheneum*).

[6] **Michel Camus** was born in 1929. He is a writer, a poet, a philosopher, a literary critic and an editor. He is also the cofounder of the journals Lettre Ouverte (1960) and L’Autre (1990). Between 1976 and 1983, he was editor in chief of the journal *Obliques*. He was director of the collection “*L’Enfer de la Bibliotheque Nationale*”, at éditions Fayard (1984–1988). He was cofounder and literary director of *Editions Lettres Vives* starting with 1981. He was a member of the board of *Centre International de Recherches et d'Études Transdisciplinaires*. He was cofounder of *Associu Paleocorsu* (Palaeolithic studies and research in Corsica). He published numerous papers in journals and collective works as well as collections of poetry. He was awarded the “Lucian Blaga” International Grand Prize in Poetry (1995). He held numerous conferences abroad: Mexico, Brazil, Quebec, Romania, Portugal, Switzerland, Belgium. Books published in Romanian: “*Parafraze eretice*”, “*Proverbele tăcerii şi ale uimirii*”, “*Imn către Lilith*”, “*Femeia dedublată*”.

The challenge of our times lies in discovering (the interaction between subject and object, an interaction which is irreducible to the subject and object—*translation*); this discovery (does not come from science nor from culture, tradition nor religion, but from the dialogue between various disciplines, enriched by knowledge—*translation of original quoted below*),[7] the work of **Stephan Lupasco**[8] being the one that revealed to Basarab Nicolescu the proximity of quantum physics in relation to transdisciplinarity.

Basarab Nicolescu advocates with convincing arguments ***a transcultural, transreligious, transpolitical and transnational attitude***[9] which is innate and can be taught *to the extent that in every human being there is an untouchable sacred core.*

The development of quantum mechanics and particle physics triggered a revolution in the history of the twentieth century. Based on these significant changes, Basarab Nicolescu in his work "*Noi, particula si Lumea*"[10] examines the nature of "reality" and **shows how the experimental study of "invisible" phenomena can lead to new concepts and even to a new world view**. The author's discourse is organized as an answer to two major questions: *What makes "the machinery of the universe" work? What can nature tell us about ourselves?* (The work reveals a surprising dimension. **Science is no longer content with manipulating beings, but now it is inseparable from philosophy; both of them must go hand in hand and regenerate human thought together**. This is a message that goes far beyond

[7] Op. cit., 19. "interacţiunii dintre subiect şi obiect, interacţiune ireductibilă şi la subiect şi la obiect", "nu vine nici dinspre ştiinţă, nici dinspre cultură, tradiţie sau religie, ci din dialogul dintre diferite discipline, îmbogăţite de cunoaştere".

[8] The Romanian-born philosopher Stephane Lupasco (1900–1988) was a bachelor in science and a doctor in philosophy at the Sorbonne. His work includes reference books in the field of philosophy, social sciences and humanities (sociology, pedagogy, psychology, psychiatry), in the field of natural sciences and exact sciences (mathematics, physics, chemistry, biology, cybernetics) as well as in the field of art. He was a nonconformist thinker in relation to the scientific and philosophical doctrines of his time, and he was a precursor of postmodernism in the field of the philosophy of science and transdisciplinarity.

[9] In Radacinile Libertatii (The Roots of Freedom), **Basarab Nicolescu**, **Michel Camus**, Curtea Veche Publishing House, Seria Actual collection, Bucharest, 2004.

[10] Having received the award of the French Academy, the book written by **Basarab Nicolescu**, an honorary member of the Romanian Academy, and titled Nous, la particule et monde, published in Romanian at Polirom Publishing House in a translation made by Vasile Sporici, was considered by Michel Camus as representing a real revolution, a key that brings together in a fundamental intuition physics and metaphysics. (We are facing a new philosophy of nature, a new Gnosis, in which the scientist's erudition is intertwined with the deepness of philosophical thinking in a fascinating search of that suspended point where the rational and the irrational come together in a sort of bet with the unknown—translation) "Ne aflam în fata unei noi filosofii a naturii, a unei noi gnoze, în care eruditia savantului se îngemaneaza cu profunzimea gândirii filosofice într-o fascinanta cautare a spatiului de gratie unde se întâlnesc rationalul si irationalul, într-un fel de pariu cu necunoscutul". The book received the prize for the best translation by the Bacau Branch of the Writers' Union in 2002.

the space of dissemination of scientific knowledge—*translation of original quoted below.*)[11] In other words, we can thus realize the importance of the study of metabolic processes, which are invisible processes, but which exist and bring about mutations in the state of the urban organism.

Being a visionary, B. Nicolescu argues with rigorousness and simplicity in favour of saving the planet and human civilization which are threatened by a type of harmful technological proliferation[12]; he was concerned about the negative impact that sciences, in their multiplicity, have on human spirituality, making thus a prediction and trying to draw a possible horizon of spiritual survival. Such a global revival would have three premises: **defining a different type of European identity** (based on the structure of Christian meditation on the Trinity, in which nature would recover its privileged role); **the successful conjunction of science and meaning** (which only means the reinterpretation according to philosophy and to the tradition of signs that nature does not cease to send); and **imagining a new mediator between science and meaning** (a new philosophy of nature which has a real ontological aperture and which is possible only thanks to transdisciplinarity).[13]

Therefore, **transdisciplinarity is considered nowadays the only model which generates unity and integration and which is able to regenerate mankind**. The importance of quantum mechanics—the science of the small infinitude—in relation to the main postmodern philosophical doctrine which presupposes the existence of only one level of reality, as well as the importance of the three levels—**microphysical and cyberspace/cybertime**, in connection with the principle of local causality and discontinuity should be noted.

We can conclude that from the point of view of classical thinking, there is nothing *between, within* and *beyond* disciplines, everything is empty, and transdisciplinarity does not exist because it has no object for study. For transdisciplinarity, though, classical thinking is not absurd, but its field of application is considered narrow.

In the presence of various levels of reality,[14] the space between disciplines and beyond disciplines is filled. Thus, the *empty void* of classical physics is replaced by the *filled void* of quantum physics, a theory which can also be found at the level of certain components of the urban landscape (for example, the road networks in

---

[11] **Edgar Morin**, "*Le Debat*", no. 40, May–September 1986.

[12] The subject is not new, and it also appeared at the **Frankfurt School** whose most important representatives are *Herbert Marcuse, Erich Fromm, Theodor W. Adorno* and *Walter Benjamin.*

[13] In "*Stiinta, sensul si evolutia. Eseu asupra lui Jakob Böhme*" ("*Science, meaning and evolution. The cosmology of Jakob Boehme*") **Basarab Nicolescu**, preface by Antoine Faivre. Translation from French by Aurelia Batali, 2nd edition, *Vitruviu* publishing house, "*Eseu*" collection, Bucharest, 2000.

[14] The various levels of reality were demonstrated with the advent of *Quantum Physics* and with the discoveries that opened a veritable Pandora's box, discoveries made by Bohr, Einstein, Pauli, Heisenberg, Dirac, Schrodinger, Born, de Broglie, etc.

the city and the streets are no longer considered empty elements—they are now important structural elements of the urban shape of the anthropic landscape).

The discontinuous structure of the levels of reality determines the **discontinuous structure of the transdisciplinary space**; this explains the complementary nature of this transdisciplinary space with disciplinary research, which, however, refers to at least one and the same level of reality or fragment thereof.

Thus, **transdisciplinarity penetrates as a study method in other** fields (biology, medicine, landscape, urbanism, land development, etc.); its task in the field of urban landscape could set the foundation for building a **conceptual system that could support any experience at the metabolic level of the urban landscape, transdisciplinarity being concerned thus with the dynamics caused by the simultaneous action of various levels of reality and establishing connections between nature, territory, cosmology and other sciences**.

The methodology of transdisciplinary research is determined by *the Levels of Reality, the Logic of the Included Third*[15] and *Complexity*, which are also the three postulates of modern science that have stood unchanged since Galileo despite the emergence of the infinite variety of methods, theories or models that have been involved in the history of different scientific disciplines.

Disciplinarity, pluridisciplinarity, ***interdisciplinarity and transdisciplinarity*** are complementary and represent the (four arrows of one and the same arc: the arc of knowledge—*translation of original quoted below*).[16] On the other hand, transdisciplinarity also refers to the idea of **scientific integration**,[17] of the great book of nature in which **each scientific discipline or art is responsible for a segment of the entirety**.

The unity of all sciences, under the sign of philosophically demonstrated religious faith, should help humanity to step into a new age of its history. Thus, a *cosmic model of globalization* in which ecumenicity and scientific reflection are intertwined is demonstrated and set forward. From the perspective of this holistic paradigm, **a new alliance between man, nature and divinity is needed**, **the first two components being the main components and the determinants of the metabolic phenomenon of the urban landscape**.

A key objective of transdisciplinarity is to develop a new philosophy of nature as a mediator of the dialogue between all fields of knowledge. The break in the

---

[15] The person who asks questions regarding what is to be found between *Why?* and *How?* between *Who?* and *What?*

[16] **Basarab Nicolescu**, *Transdisciplinaritatea. Manifest.* (Transdisciplinarity. A Manifesto.), *Polirom* Publishing House, *Idei Contemporane* Collection, *Plural* Series, Iasi, 1999, 5—"*patru săgeţi ale unuia şi aceluiaşi arc: cel al cunoaşterii*".

[17] For **Basarab Nicolescu**, *Joc Secund,* the book of poetry written by **Ion Barbu** (the pseudonym of the mathematician Dan Barbilian who has an international reputation), which was published in 1930 and which generated critical interpretations from specialists in such various field such as mathematics, alchemy and Christian mysticism is a (jewel of transdisciplinarity—*translation*) ("*giuvaer al transdisciplinarităţii*") and **Eminescu-Blaga-Barbu** represent (a fundamental axis of Romanian culture—*translation)* ("*axă fundamentală a culturii românesti*").

field of science, caused by Einstein (through the *Theory of Relativity*), by Heisenberg (through the *Principle of Indeterminacy*) and by Max Planck (through the *Quantum Theory*), led to the collapse of traditional paradigms, thus demonstrating that there are no definitive theories nor exclusive systems and that the human being itself has remained an enigmatic issue. (In the macrocosm, Einstein's **Theory of Relativity** showed that the universe is not populated by Newtonian objects but by infinite vibratory phenomena, that space and time form a four-dimensional continuum, that space is not three-dimensional and time is not one-dimensional—*translation of original quoted below*)[18]

Quantum physics has highlighted the phenomenon through which **the process of measuring modifies the studied phenomenon**—in other words, ***the researcher is never independent or neutral in relation to his investigation***. By extrapolating this situation and extending it at the level of the whole universe, it means that all the aspects of the universe are interdependent; thus, we have the holistic vision validated by various theories, a vision which is characteristic of man, nature and society which are important components of the metabolism of the urban landscape.

In the first half of the twentieth century, **the transpersonal paradigm** appeared which capitalized worldwide perennial philosophy, modern physics and current research on consciousness. The impressive amount of data collected through direct experience, under the aegis of transpersonal orientation, led to disturbing truths. It was thus proved that (the representation of reality is dependent on the state of consciousness in which we find ourselves. As a result, the fourth state, in other words, the superconsciousness reveals an ineffable reality that is yet unknown to scientific classicism and which opens a new window towards the hidden parts of the world—*translation of original quoted below*).[19] In this sense, the representation and the sensitive reading of the reality surrounding the natural and/or the urban landscape is dependent on the state of the consciousness of the urban actor involved in the process. By using the superconsciousness, we can reveal through research and sensitivity new windows in the study of landscape, which may lead to different results, study methods and methodologies of the natural environment and of the urban man-made environment.

In early human history, science and culture were inseparable, having the same principles on the meaning of the universe and of life. During the Renaissance, this link had

---

[18] **Anca Munteanu**, *Pledoarie pentru o noua arhitectura a conştiinţei* (Plea for a new architecture of consciousness), online Journal of the Romanian *Association for Transpersonal Psychology*, No. 1–2/2004. (*În macrocosmos,* ***Teoria Relativităţii*** *a lui Einstein, a demonstrat că universul nu e populat de obiecte newtoniene, ci de infinite fenomene vibratorii, că spaţiul şi timpul formează un continuum cvadridimensional, că spaţiul nu e tridimensional, iar timpul nu e unidimensional.*)

[19] **Mânzat I.,** Psihologia transpersonala (Transpersonal Psychology), Iasi, Cantes publishing house, 2002; "reprezentarea realităţii este dependentă de starea conştiinţei în care ne aflăm. Ca urmare, cea de-a patra stare, adică supraconştiinţa ne dezvăluie o realitate inefabilă şi înca necunoscută clasicismului ştiinţific, deschizând o altă fereastră spre ascunderile lumii".

not been broken yet, the first *University*, as its name indicated, studied *the universal*, which existed in those who, regardless of the field of their work, have left their mark on the history of knowledge.[20] The founders of modern science, however, have nothing in common with the stereotypical image of the Renaissance man of science; a break has taken place between science and culture that no longer share anything in common, as well as a fragmentation within one and the same culture and science.[21] The separation *Science/Culture* caused the myth of the separation between the West and the East: the West as the depository of *science as a way of understanding Nature* and the East as the depository of *wisdom as a way of understanding the human being*, as well as the separation between nature/natural landscape and anthropic/urban landscape. Lately, there has been a rapprochement between science and art but also between the natural landscape and the anthropic landscape due to the explosion of information but also due to the cultural landscape. A new type of landscape is born nowadays by means of transferring informatics methods,[22] which reflects new aspects of *information*, of creating new forms, forms that are ever-changing, arising from the collective imagination of landscape artists, urban planners and architects.[23]

Recognizing the existence of different levels of reality governed by different logics is inherent to transdisciplinary attitude and to landscape studies (in all the components of landscape: natural, anthropic and cultural) and to urban research and land planning. Any attempt to reduce reality to a single level governed by a single logic is incompatible with transdisciplinarity.[24]

The type of transdisciplinarity used in landscape research is complementary to disciplinary approach by confronting various disciplines considered conventional (urbanism, land planning, geography, architecture, art, history, horticulture, geology, forestry, anthropology, ethnography, sociology, the psychology of the built space, planning and management, etc.) but also by confronting disciplines considered unconventional[25] (such as quantum physics, complexity, computer science,

---

[20] The inventor of imaginary numbers, Cardan, was also the inventor of the suspension system that bears his name and he was also a mathematician, a doctor and an astrologer. Kepler was an astronomer and astrologer. Newton was a physicist, a theologian and an alchemist who was passionate about the Trinity but also about geometry.

[21] For instance, in science, there is a distinction between exact sciences and humanities or within hard sciences and soft sciences—the terms used in English terminology through which an exact denomination and the antagonistic perception of the two was sought.

[22] *Computer art* becomes spectacular by using the information that travels through *Internet* as new material.

[23] There are manifestations through which the landscape is seen as an unconventional art form in which landscapers, architects, artists and scientists from various fields meet in cyberspace (on the Internet) to create together using sounds, colours and images.

[24] See *Article 2* from the *Charter of Transdisciplinarity.*

[25] See the unconventional methods promoted in the book written by Cerasella Crăciun "*Metabolismul urban. O abordare Neconventionala a Organismului Urban*" (The Urban Metabolism. An Unconventional Approach to the Urban Organism) (366 A5 pages in colour), "Ion Mincu" University Publishing House, ISBN 978-973-1884-14-1, Bucharest, 2008 (a work selected by an international jury at BAB—ARCHITECTURE BIENNALE Bucharest, 2008, *Publications* section).

linguistics, music, geometry and sacred geography); thus, transdisciplinarity brings new results, builds new bridges between disciplines and offers a new vision on nature and reality. Transdisciplinarity in landscape study does not seek to create a larger discipline meant to encompass all other disciplines, but it seeks to **open all disciplines to what they have in common and to what lies beyond their borders**.

The key to transdisciplinarity is the semantic unification of the meanings *between* and *beyond* various disciplines that research and influence natural, urban and cultural landscape. This unification requires a rationality that is opened by a new vision on the relativity of the concepts of "definition" and "objectivity" in landscape, from macroterritorial landscape, to mezzo-landscape and detailed landscape, from human settlements, to sensitive details and inserts in landscape.

The transdisciplinary vision in landscape is *open* to the extent that it exceeds the field of natural sciences (mathematics, physics, topography, geography, geology, horticulture, etc.), but especially the field of the sciences of synthesis (such as landscaping, urbanism, architecture, anthropology, etc.) that are on the edge between science and art, through the reconciliation and the dialogue established not only with the humanities but also with the arts, with literature, poetry, inner space and experience. Thus, the relationship between science and philosophical, religious or spiritual concerns is not just a concern of the Renaissance; this relationship becomes a topic of increasingly heated debate nowadays, including in the field of natural, cultural and anthropic landscape.

We can conclude that progress is necessary for spiritual discoveries as well as for other dimensions of experience and human endeavours, including in the field of landscaping, urbanism and architecture in relation to the study of the city seen as an urban organism but also as a processor of human life. The same concerns regarding the connections between different branches of science, concerns that are the result of the need for transdisciplinarity, make their presence felt even in the field of education.[26] The interdisciplinary perspective involves addressing complex phenomena and processes at the level of the landscape and of the urban

[26] Education (has the difficult task of transmitting a culture accumulated for centuries but also of training subjects for a largely unpredictable future—translation) ("are dificila misiune de a transmite o cultură acumulată de secole, dar şi o pregătire pentru un viitor în bună măsură imprevizibil") as Jacques Delors maintained in Comoara lăuntrică. Raportul catre UNESCO al Comisiei Internaţionale pentru educaţie în sec. XXI (The Treasure Within. Report to UNESCO of the International Commission on Education for the twenty-first century) Polirom, Iaşi, 2000. Also, in this context, Professor Dr. Vasile Marcu and Lecturer Dr. Mariana Marinescu from the University of Oradea, at the course "Implementarea tehnologiilor in educaţie sau educaţia tehnologica" (Implementing technology in education and technological education), aim at discussing the directions to be taken in restructuring educational realities at the beginning of this millennium in a society characterized by the accelerated dynamics of structural changes, but also by the depth of these changes and by placing the human being in the context of these "technological outbursts" in which the way mankind will take is unclear: towards globalization or towards atomization. In the context of the explosive evolution that has taken place across all the fields in the last 10–15 years, especially in those fields that have determined a problematic change, technological education is inserted and is considered a metamorphosis of traditional education mixed with the principles of the new educational paradigm, including here landscape, urban and architectural education.

environment, contributing to the formation of a *sensitive image in a landscape* and to the *acquisition of a unified methodology of approaching urban realities*; the difference between the interdisciplinary perspective and the pluridisciplinary perspective lies in the fact that in the case of the latter, the perspectives, phenomena and processes seen at the level of the natural environment and of human settlements are the object of study of various disciplines which contribute to a better understanding of the processes and phenomena, each discipline with its own methodology. Understanding complex phenomena and processes that take place in the natural, anthropic and cultural landscape is achieved within border disciplines from different but compatible angles. The transdisciplinary approach refers to exploring reality and urban phenomena and to the exercise of very complex landscape processes so that by coordinating researches, we are able to define new areas of knowledge.

The Delors Report emphasizes **four pillars of a new type of education**: ***learning to know***,[27] ***learning to do***,[28] ***learning to live with the others***[29] ***and learning to exist***.[30] In this context, the transdisciplinary approach can bring an important contribution to the establishment of a new type of education and this approach will allow the evolution towards the forgotten mission of education—*studying the universal*, so that universities can become the (focal point of a new type of humanism—*translation of original quoted below*).[31]

From the confrontation between disciplines of the complex phenomena and processes, transdisciplinarity creates new points of intersection between disciplines and new results and offers us a new vision and interpretation of the specialized education in the field of landscaping, urbanism and architecture (and not only), as well as a new approach to understanding the world, culture, man and society, a new vision on nature and on the realities of landscape because (the scientific Spirit unites mankind at an abstract and rational level but it is precisely this unity that gives human civilization its universality—*translation of original quoted below*).[32]

---

[27] This refers to learning the methods that help us to distinguish between what is real and what is an illusion and to have an access way to the knowledge of the era we live in.

[28] This refers to acquiring a job and the knowledge and practices associated with this job, as well as learning to be creative by avoiding the danger of the specialty.

[29] This refers to complying with the rules that regulate the relationship between the human beings that form a community which is open to transcultural, transreligious, transpolitical and transnational attitudes.

[30] This refers to exploring our certainties, beliefs, conditionings, to a type of (integral education of man—*translation*) "*educație integrala a omului*" (Rene Daumal).

[31] A type of humanism which **Basarab Nicolescu** calls *Transhumanism*, in *Transdisciplinaritatea. Manifest.* (Transdisciplinarity. A Manifesto.), *Polirom* Publishing House, *Idei Contemporane* Collection, *Plural* Series, Iasi, 1999, 164—"*focarul unui nou tip de umanism*".

[32] **Mircea Malița** (mathematician, essayist, scholar, diplomat—ambassador in the USA and Director of the Romanian Library in New York, university professor) in *Zece mii de culturi o singura civilizație* (Ten thousand cultures, one civilization), *Nemira* publishing house, Bucharest, 2002—"*Spiritul științific unifică omenirea la un nivel abstract și rațional, dar tocmai aceasta dă civilizației umane un caracter universal*".

References

Alexander C (1987) A new theory of urban design, The MIT Press
Bejan A (2000) Shape and structure, from engineering to nature, Cambridge University Press, Cambridge, MA
Bergson H (1911) Matter and memory. Allen and Unwin, London
Bernea E (1997) Spaţiu, timp şi cauzalitate la poporul român (Space, time and causality for the Romanian people). Humanitas Publishing House, Bucharest
Blaga L (1968) Despre conştiinţa filosofică (On the philosophical consciousness). Facla Publishing House, Timisoara
Bohm D (1974) Science as perception-communication. In: Suppe F (ed) The structure of scientific theories. University of Illinois Press, Urbana
Bohr N (1998) Atomic physics and human knowledge. Wiley, New York
Capra F (2004) Momentul adevărului: ştiinţa, societatea şi noua cultură (The moment of truth: science, society and new culture). Tehnica Publishing House, Bucharest
Castells M (1996) The rise of the network society. Blackwell, Cambridge
Choay F (1997) The rule and the model: on the theory of architecture and urbanism, The MIT Press
Craciun C (2008) Metabolismul urban. O abordare neconvenţională a organismului urban (The urban metabolism. An unconventional approach to the urban organism). "Ion Mincu" University Publishing House, Bucharest
Delors J (2000) Comoara lăuntrică. Raportul către UNESCO al comisiei internaţionale pentru educaţie în sec. XXI (The treasure within. report to UNESCO of the international commission on education for the twenty-first century). Polirom Publishing House, Iasi
Dogan M, Pahre R (1993) Noile stiinte sociale. Interpenetrarea disciplinelor (The new social sciences: the interpenetration of disciplines). Academiei Romane Publishing House, Bucharest
Draganescu M (1989) Inelul lumii materiale (The ring of the material world). Stiintifica si Enciclopedica Publishing House, Bucuresti
Driesch EG (1929) Science and philosophy of the organism. A and C Black, London
Eliade M (1992) Sacrul şi profanul (The sacred and the profane). Humanitas Publishing House, Bucharest
Heisenberg W (1963) Physics and philosophy. Allen and Unwin, London
Hall P (1998) Cities in civilization. Pantheon Books, New York
Hawking S (1997) Visul lui einstein şi alte eseuri (Einstein's dream and other essays). Humanitas Publishing House, Bucharest
Heidegger M (1994) Fiintă şi timp (Being and time). Jurnalul Literar Publishing House, Bucharest
Jung CG (2007) Despre fenomenul spiritului în artă şi ştiinţă (On the phenomenon of spirit in art and science). Trei Publishing House, Bucharest
Lupasco S (1982) Logica dinamica a contradictoriului (The dynamic logic of contradiction). Politica Publishing House, Bucharest
Lupasco S (1992) Experienţa microfizică şi gândirea umană (Microphysical Experience and human thought). Stiintifica Publishing House, Bucharest
Lupasco S (1979) L'universe psychique, Denoel-Gonthier, col. Meditations, Paris
Lupasco S (1986) L'energie et la matiere psychique. Julliard, Paris
Lynch K (1998) L'image de la cite. Dunod, Paris
Koestler A Lunaticii. Evoluţia concepţiei despre Univers de la Pitagora la Newton (The sleepwalkers: a history of man's changing vision of the universe). Istoria Ideilor collection, Humanitas Publishing House, Bucuresti
Manzat I (1999) Psihologia sinergetică (Synergistic psychology). Pro Humanitate Publishing House, Bucharest
Manzat I (2002) Psihologia transpersonală (Transpersonal psychology). Cantes Publishing House, Iasi
Moles A (1972) Psychologie de l'espace. Casterman, MO

Moraru I (1992) Strategii creative transdisciplinare. Introducere în scientoeuristică (Transdisciplinary creative strategies. Introduction to scientoeuristics). Academiei Romane Publishing House
Morin E (1986) Le Debat, no. 40, mai–septembrie
Mumford L (1961) The city in history: its origins, its transformations, and its prospects. Brace and Co., New York
Munteanu A (2004) Pledoarie pentru o nouă arhitectură a conştiinţei (Plea for a new architecture of consciousness), online J Romanian Assoc Transpersonal Psychol, No. 1–2/2004
Nicolescu B, Camus M (2004) Rădăcinile libertăţii (The roots of freedom). Curtea Veche Publishing House, Seria Actual Collection, Bucharest
Nicolescu B (2000) Ştiinţa, sensul şi evoluţia. Eseu asupra lui Jakob Böhme (Science, meaning and evolution. The cosmology of Jakob Boehme), 1st Edition, Vitruviu Publishing House, Eseu Collection, Bucharest
Nicolescu B (1999) Transdisciplinaritatea. Manifest (Transdisciplinarity. A Manifesto.), Idei Contemporane Collection, Polirom Publishing House, Plural Series, Iasi
Noica C (1978) Spiritul românesc în cumpătul vremii. Şase maladii ale spiritului contemporan (Six maladies of the contemporary spirit. The Romanian spirit at the conjuncture of time). Univers Publishing House, Bucharest
Norberg-Schultz C (1986) Genius loci. Pierre Mardaga
Norberg-Schultz C (1971) Existence space and architecture. Studio Vista, London
Odobleja S (1982) Psihologia consonantistă (Consonantist psychology). Stiintifica si Enciclopedica Publishing House, Bucharest
Prigogine I and Stengers I (1984) Noua Alianţă. Metamorfoza Ştiinţei. (The New Alliance. The Metamorphosis of Science). Politica Publishing House, Bucharest
Steiner R (2005) Metamorfoză cosmică şi umană (Cosmic and human metamorphosis)., Biblioteca Antroposofica SeriesUnivers Enciclopedic Publishing House, Bucharest
Steiner R (1999) Introduceri la scrierile de ştiinţe naturale ale lui Goethe (Introduction to Goethe's scientific writings). Triade Publishing House, Cluj Napoca
Stengers I (2001) Inventarea ştiinţelor moderne (The invention of modern science)., Plural M CollectionPolirom Publishing House, Iasi
The Charter of Transdisciplinarity, Convento da Arrábida, 6th of November 1994, Drafting Committee: Lima de Freitas, Edgar Morin and Basarab Nicolescu (translation into Romanian by Horia Vasilescu)
Transdisciplinaritatea. Manifest. (Transdisciplinarity. A Manifesto.), Idei Contemporane Collection. Plural Series, Polirom Publishing House, Iasi, 1999

Part II
Archetypal landscape: Philosophy/Epistemology/ Semiotics—Specificity

Living the Space from Ţara Haţegului: Building Places and Landscapes as Collective Identity and Memory

Mihaela Hărmănescu

Abstract The communication refers to the active role that the space has in the building of cultural schemes, seen as the memory and identity of a place. The intention is to identify types of places with the possibility of overlapping and communication between them within the territory very well defines, Ţara Haţegului. The development of Ţara Haţegului as the GeoPark leads to the development of some publicly and semi-publicly differentiated spaces within the settlements, which means, in fact, the relationship between the different types of landscapes at the scale of human settlement. The spatial analysis from the morphological point of view highlights the interaction between the geographical configurations and the evolution of the society's culture over time within the Romanian rural landscape.

Keywords Culture • Human action • Settlement • Rural • Vernacular

1 Introduction

Halbwachs (1997/1950) believes that every society cuts out the space (…), so that it a fixed frame in which it closes inside to find its memories, highlighting the thresholds between the spaces where different behaviour rules reign, referring to different memories. The human settlements provide a type of space hierarchy, directly resulted from the identity and memory process of the society: areas identified within the territory by a cultural approach are established in spaces of collective cultural memory, milestones for the society in the evolutionary approach of the GeoPark. What is important is the impact that the revitalization of these spaces manifest over the localities. The rules of the fragmentation mode, the type of basic units, their occupation way, use, can be defined as memories of the place, and they contribute to the specific compliance of

M. Hărmănescu (✉)
Faculty of Urban Planning, Department of Urban and Landscape Design, Ion Mincu University of Architecture and Urbanism, Academiei 18-20, 010014 Bucharest, Romania
e-mail: mihaela.harmanescu@gmail.com

C. Crăciun and M. Bostenaru Dan (eds.), *Planning and Designing Sustainable and Resilient Landscapes*, Springer Geography, DOI: 10.1007/978-94-017-8536-5_2,

the rural organism. Exploring viable alternatives to the actual and future relationships of the citizens of the rural public space offer it an active role in the developmental approach of a touristic potential according to the principles of sustainable development. Identifying the cultural, architectural vernacular of heritage values, the landscapes and the relationships in which they communicate and overlap, constitutes an important approach in the study dedicated to the settlements of Ţara Haţegului.

2 The Concept of GeoPark

Since the United Nations Conference on Environment and Development, held in Rio de Janeiro in 1992, not only the policy makers, the scientists but also the public understood that the protection and the wise management of the environment are a priority. It is known that, to respect the environment, requires a better understanding of the geological, biological and physical processes which left their mark on the earth. A good understanding of the geological heritage and a healthy respect for what it represents are the important factors in the holistic approach and for a sustainable development. In support of these ideas, there have been geo-conservation international initiatives suggested and launched, counting that "geo" is more than geology; it involves geographical and geomorphological issues as well (Farsani et al. 2012). The GeoPark, concept created by UNESCO 1990, is defined as a "territory which includes an outstanding geological heritage and a territorial strategy of sustainable development supported by a European programme. It has, of course, well-defined borders and a surface which is enough for a real economical and territorial development". The majority of the sites on the territory of a GeoPark require a certain number of geological sites, but their interest can be also archaeological, ecological, historical, ethnological or cultural, strengthening of the cultural, social identity, of protection and the natural or cultural heritage. A GeoPark has an active role in the economical development of the territory or through the improving of a general image about the geological legacy and the development of the geo-tourism. The focus of this article is not on the specific characteristics of a GeoPark, but on a Romanian territory that corresponds perfectly with the requirements of UNESCO: Ţara Haţegului. Its mention as GeoPark is important because together with its promotion and development, it will attract different types of tourists and investors accordingly, requiring a special attention on the development of the settlements which belong to this territory.

3 Țara Hațegului: Territory Between Preservation, Development and Promotion

Situated on the main road which tied, even since antiquity, and then in the middle ages, Transylvania to Banat, with the custom from Bretea, Haţeg and Poarta de Fier, and also which had one of the most notable hearths of civilization from the whole Romanian space.

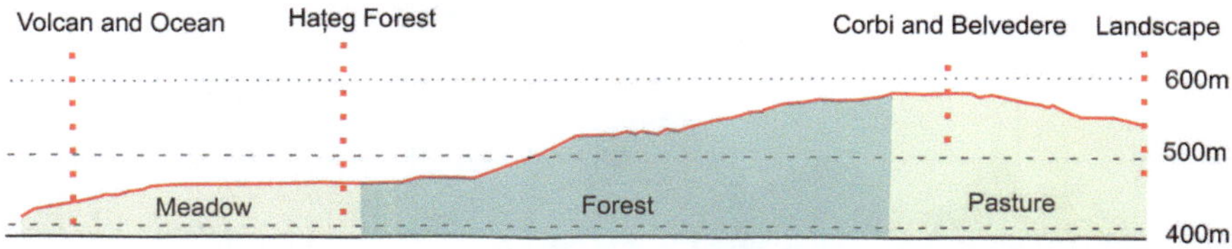

Fig. 1 Macro section—Denus-Stei volcanic landscape. From the point of view of land use, the largest surfaces are covered by forests, grasslands have a percentage slightly reduced, and the most under-represented land is the arable one (*source* the author)

The varied relief, consisting of field watered by the Strei River, Great River and the Yellow River and their numerous tributaries, from the piedmont and hills covered with forests and rich grasslands were the foundation of a thriving mixed economy, agrarian pastoral. The fishing, hunting, gold washing from the rivers, exploitation of iron, limestone, marble, complement the traditional occupations of the inhabitants.

The sense, which the concept of "country" has acquired over time, characteristically the territory in question, is the mental space one, territorial entity (regional area) "which a community but also an individual belonging to it, integrates it its/his own scale of existent values through perception (l'espace perçu), experience (l'espace vecu) and imagination ('imagined space')" (Lefebvre 1974). The regional entity emerges through three major significances: the territory one, political-administrative unit and system. The analysis of the regional space of Ţara Haţegului is the most important step in restoring the local–regional identities, the importance of studying it resulting from the fact that the region is considered to be the random circumstance within which the people who reside it formed, in which they act like agents and which components, structures, represent the ways through which people organize their lives. Like any other region, the territorial structure has a role in the implementation of the sustainable process of development, at the same time being also a cultural entity for the local communities which create the process of economical competitiveness and influence the modernization. The landscape, through its specificity, highlights the mountains with extensive fields, at high height for sheep, the fir, beech and oak tree forests down to the hills; the valleys that open to fields and villages (Fig. 1).

Therefore, the GeoPark project, considered 21st Local Agenda of the territory, cements the general frame for a sustainable development, and it values the cultural landscape if Ţara Haţegului, distinct territory by its folkloric and traditional crafts, unmistakable popular customs, popular traditions and customs very well preserved.

The identification of the varied landscape is done starting from the last group of dinosaurs, on one hand, to the settlements of historical, scientific and cultural importance, which covers a period of 4,000 years and which, nowadays, consists of milestones for the community of Ţara Haţegului. This territory contains well-preserved ruins: amphitheatres and temples of the Roman capital of Dacia, medieval fortresses, castles and forts since thirteenth century, churches

Fig. 2 The Dacian Costesti Citadel 101–102 i.Hr (*source* author)

and monasteries, memory and identity places. Today, as the signs of time, these memory places carry the past in our daily lives, and they become the base for our cultural collective memory. From the man who lived in caves (Ohaba-Ponor, Cioclovina), where there were found polished objects from the Dacian era and kept at Blidaru, Costeşti and Gradiştea Muncelului (Fig. 2), continuing with the ethno genesis of the Romanian people (Daco-Romans), to the old socio-political compositions of the Middle Ages, every era put its mark on this ancient Romanian hearth.

The first Documentary evidence of Ţara Haţegului dates from the medieval, in Diploma Cavalerilor Ioaniţi in 1247, with the name Terra Haszok. The further dwelling after this date has been proved through a series of written documents or through buildings, which is a true heritage, which is not fully exploited. The origin of the name of Ţara Haţegului has received various explanations and interpretations; the name might come either from Hungarian-hat-zeg, having the meaning of fang or peak being located in the back, on the shoulder (on the hill), either from a Romanian anthroponomy, or it is related to the nickname "haţeg" ("tiny forest, young, bush, thicket").

Ţara Haţegului was recognized the outstanding place it occupies in terms of age and large number of Romanian medieval monuments it contains. It is about 20 monuments, dated prior to the fifteenth century, hosting numerous ensembles of mural paintings from fourteenth to fifteenth centuries. From this land comes the most important family in the history of the Hungarian Kingdom from the fifteenth century—Corvinesti—and one of the brightest personalities of the history of the south-eastern Europe—Iancu de Hunedoara. Moreover, over the past of Ţara Haţegului have stopped great Romanian historians Nicolae Iorga, Bogdan Petriceicu-Haşdeu, A. D. Xenopol, Nicolae and Ovid Densusianu, ethnologists, researchers and so on.

What we call today Ţara Haţegului does not correspond in dimension with the ancient Romanian territory with the same name, which was part of the principality

Fig. 3 **a** The Densuş church. **b** Santamaria Orlea church (*source* author)

of Litovoi, as certified by the famous Diploma Cavalerilor Ioaniti from 2nd June 1247, by which the Ioaniti Knights receive parts from the country of Litovoi, except the country of "Harzszoc". This domain fit into a much wider one, which later became Ţara Românească, as it was kept in the people's consciousness that Negru Vodă was born in Haţeg from where he went to south.

Synthetic, compared to other European spaces, one can observe the remaining of some ways of living preserved even today. In response, the last decades represent a time of acceleration of the dissolution process of the landscape's value, of human groups and of the societies, the traditional and rural cultures almost on the entire country, and the common solutions of protection, rehabilitation, promotion, until now they are at a superficial level of the phenomenon.

For Ţara Haţegului, the revaluation, as a potential and resource of the natural and cultural heritage, unique in the country by diversity and age, represents, together with the local human resources, the main opportunity of developing of the area, the territory being well defines in space, both through geographical characteristics and the very specific cultural and historical traditions. This geographical space of Ţara Haţegului appears as a document, written and re-written (palimpsest) through the people from all the eras, better said as a result of the different temporalities, incompletely preserved. This space, integrated in a project on an international scale, transforms the landscape and attracts, at the same time, a major interest, and it represents an important resource to attract visitors, because tourism is essential in this region. The most important part of the existing activity in the territory consists of the variety of archaeologically, historically and culturally relevant sites. These sites include the well-preserved ruins, including the amphitheatres and temples, of the Roman capital Dacia, medieval fortresses, castles and security towers since thirteenth century, churches and monasteries, including the old churches from Densuş and Santamaria Orlea (Fig. 3).

The territory, which by definition is characterized by individuality and specificity determined by the environment conditions, on one hand, and by the economical, social, ethnological and cultural conditions on the other hand, raises a series of issues related to the relationship space-community: the way the society transforms the space it occupies, within the settlements, especially in the ones near sites of archaeological, historical and cultural interest, resources for attracting visitors. It seems that, in general, there is really difficult to understand that a certain territory, but shaped prior to the modern era (we could say traditional) should not necessarily be protected as a memory place or historical value or environmental or stylistic, or as a touristic object, but through which it contains as an intrinsic architectural, town planning, landscape value: conformation, materials, technology. Without this way of thinking, the settlements are subject to a continuous pressure of the dynamics of the human evolution, pressure which induces the society of those places "states of tolerance" (Choay 1998).

How will the community from each of these settlements the modelling of the space of Ţara Haţegului and how will it materialize it? How will it highlight the identity of each settlement? The reason is a space apparently common can offer now an important support in the possibility to form alternative models in the social building of the landscape of Ţara Haţegului. Moreover, the space is used as a filter and as a model of the human behaviours, being a component of knowledge over time.

4 Dwelling of the Hațegan Village: Tradition and Landscape

The landscape, in general, is since some time reported to the national and local policies and to the community spirit and each community (or individual) (de)limits within the landscape a centralized space defined by himself as his own place, a place (of the man) which from the formal point of view, Norberg Schulz (1986) considers it round, meaning it has a centre in which there is the man or the community. The problem of the identity, of the place, ties the phenomenon tradition to the environment in which it was born and was shaped over time and to solve the problem of the identity of a place.

The Romanian Haţegan villages, as the materialization of these spaces, are, indeed, a gathering (grouping) of houses (more or less dispersed) consistent with a spatial symbolism, directly related to the place and highlighting the eternity based on a mentality of the celestial and organic, which centre is occupied is assimilated to the mundi axis and it is marked by "the stake fixed in the middle of the village" (Creţu 1988) (Fig. 4).

The village is the fundamental form in the social life (Gusti 1974), which is also noticeable in its spatial development: organic or linear along a road that links it to another site, without having a pre-established plan. Although located on a lower level of organization of the human habitat, it is characterized by a complex feature defined by the presence of a well chosen hearth, in which there are concentrated the homes and dependencies of the farmers, as well as by a "dust", the work

Fig. 4 Typologies of rural settlements in Ţara Haţegului (*source* author)

place found outside, the duties and the space of the productive activities of the people (Cucu 1981, quote by Vert 2002). The village has its origins in prehistory, the appearance as human groups. It has a continuous inside with different typologies: some geometric rectangular areas, areas organized around some markets, then irregular and deformed areas, as a result of the morphology of the landscape and infrastructure.

There can be noticed conclusive differences between the initial systems of developing, often cellular, while the later system, upgraded, has a more clear geometry, despite the local deformations. What is truly remarkable is the stability of the factors that influence the reality of the settlement and of the community today, due to some systems of filtration of the factors: the organization itself, of the built environment, corresponds to this concept; it allows or denies (Gheorghiu 2008).

Relating to the dates concerning the genesis and morphological evolution of the village, the logic of the spatial organization of the localities is even more difficult to deduce. In his study on the villages from Haţeg (Conea 1935), at the beginning of the twentieth century, he considers that there are no significant landmarks, as in the case of the centre, almost unexpressed from the spatial point of view and expressed only in some degree from the functional point of view. ≪"the Centre" designated by the residents is a fragment of the street, between an intersection—the one with restaurants and school, along with which there is also the cultural community. There are two churches in the village, but by the location and by its common architecture, they do not stand out as landmarks (although both are listed on the historical monuments' list.≫

Although the village is not characterized by a central space, there can be noticed though, interest points, places expressed from the point of view of the space, with a social value.

They are different, depending on their level of emancipation and community development, whose appreciation is often a subjective matter, because it is related to the reference system of the observer. Worth mentioning, here, is that the way of life of the traditional society was based almost exclusively on decisions following practical utility, the external stimuli receiving an answer in full compliance with daily or long term needs, hence the certain hierarchies in the filled of conceptual and material reality of the group life. Practices, habits and customs were more important than their realization at some point in time. That is, in the case of the Romanian village there is no centre, but the place is marked by landmark and related to the road axis, where the daily life of the community is developing. The lack of a place (mark) is a model in the individual collective consciousness of the existential space, it has an identity linked to tradition and it is built over time, as a mirror of the society and the individual's experience.

Creating (unique) spaces is a natural, natural and inalienable gesture that is developed within a morphological framework defined by social factors (the cultural model and through tradition system). Thus, anyone can create a place within a morphological framework defined by anthropogenic models (tradition, culture, etc.) and based on a physical support. Once created, the place can be modified and enriched by successive accumulations of human experiences over generations, and it can be opened more than any other anonymous and rigid place. The place, as public space becomes the public's space, where it participates, spontaneous and legitimate at the same time on the establishment of the landscape (Fig. 5).

Community, the one that creates this landscape, brings with it also the problem of the identity and belonging to a group and thereby the inhabited space and therefore "landscaped" (Tudora 2006) by this group. We are here in front of a concept that balances between the physical, aesthetic and social space, which is defined as a designed space, lived space, perceived space (Lefebvre 1974). After Lefebvre, the designed space is based on a series of representations of the space and conventions (scientists and planners). The living space is a space consisting of images and symbols that accompany the residents or the artists of the performances. The perceived space is intimately related to the daily life (use of time, travelling, places, etc.).

The landscape is a place of conflict, the place of a perpetual compromise (…) (Jackson 1984), which we can consider as a sum of these three spaces. "It is designed because it started from the structure of the collective identity. It is also

Fig. 5 Rural public spaces (*source* author)

perceived, over the customary practices (and through its vernacularism), and it is experienced as it is symbolized; it bears within itself the daily and power signs".

In reality, these landscapes are not so strong anymore, the village lost its image over time, and they became an "open work" being characterized by a continuous dynamic, due to human actions (economical, social and cultural factors) and nature's actions (bio-physiological processes).

The tradition that we face today contains stages and components of the town, and its relation with the modernity is not necessarily also decisively dramatic and it is plausible to be able to continue in a peaceful manner, even by losing lately the cohesion of the rural communities, carriers of tradition. The answer, customized to the reality of the area, highlights the disintegration of the community and the way to act, the tradition not longer belongs to this community. The signs of behavioural mutations belong to various historical periods, interventions from outside, continued and continue to grind the internal structures of the rural community, being constantly by chance in front of the inherited landscape: it re-shapes it, it builds it, it modernizes it or uses it the way the know.

With the disappearance of the traditions old jobs have disappeared too, as an important component of the village, and the new constructions in the countryside, new tourist boarding houses, instead of using the traditional landmarks they use modern inappropriate landmarks. This created a different type of landscape, a popular one, formed by mass practice, with an unpleasant visual impact but also a destruction of the local identity took place: perpetual construction and the indifference of the official aesthetics that shapes the daily space (in situ vernacular building by Jackson) emphasized by a visual landscape, which it is created also out of this trend of modernization. Existing communities, resulted from a social and economic dynamic, have lost everything they gained over time in the field of the collective mentality, without gaining anything that a reverse of values and a bizarre reference to the modernity.

5 Place Architecture: Reference and Identity

Anyone can participate spontaneously on the landscape construction, in the absence of a coherent project which would aim for the landscape construction, as a perceived and enriched space. They are those who, by making the space vernacular they produce transformations, both in situ and in visu.

In situ through the direct effects on the spatial morphology and in visu through affirmation and as a reading grid of the space: a landscape of several individual or collective models that work alongside the official models. And all this because the place's image reveals, also, the daily existence of the individual, which is the connection with his life frame: the material and emotional frame, realizing thus a specific or that spirit of the place "genius loci" (Lobbel 1979 quote by Machedon 2006).

It is controlled by a system of symbols that expresses the character of spatial relations (unit-all, individual-landscape).

The analysis of the places as physical objects in terms of distinguishing characteristics (specificity) is the necessary knowledge underlying the grounding of every intention to upgrade the identity and memory, in accordance with the principals of the European convention (identification) and according to the various cognitive methodologies developed by the European countries.

From the physical point of view, the landscape is "architecture" (in a broad sense this could be read as architecture) and, as such, it is a manufact (manufactus), characterized by the specific natural or anthropological elements/materials and construction techniques.

Thus, the places have a three-dimensional organization, are spaces made of elements that define them, spaces made of construction materials and techniques (to reinforce the concept the term manufact was introduced to emphasize the relevance of the places' characteristics, this prompting their specificity).

The image of the place obtained is evolving and has a tendency to constantly adapt to the successive changes occurring in the relation of approaching, using, amending and consecration (often divine) of the space.

What is proposed here is to retrieve the lost identity of these villages, because the GeoPark develops the geo-tourism, or the agricultural tourism is poorly understood by the community; it is promoted everything that exists as memory and everything it is possible to distinguish as architectural, economical, social, ethnological and cultural value. The intervention is materialized in identity projects which differ on each village, in comparison with many other parameters: position, potential, history, heritage included.

Taking as an example Sarmizegetusa Ulpia Traiana a small village, but, with a Roman archaeological site of great historical value, it is noticed that the tourists stop in the village only to visit the archaeological site, without having any interest in the remarkable architecture in stone and wood or the traditions of the place. The problem of continuity of a certain spirit of tradition can be seen both in terms of historical and cultural (that claims the preservation of some elements of the tradition, history and national culture for memorial, sentimental, cultural, educational reasons or others) as well as from a pragmatic point of view, through which the tradition may be included in the daily process of promoting the living. In view of preserving the collective memory, it was included in the site's documents, one of the chances being the "storage" of some elements in the museum or "in situ".

6 Collective Memory: Cultural Landscape

The recovery of the tradition means to recover a natural way of living, in relation to the natural rhythms and the place's characteristics. The problem, in both cases, is the compatibility: to what extent and which of the components of the traditional housing are able to satisfy the needs of the moment and to lead to the shaping of a creation of a modern approach of the landscape, in other words, through the spirit of tradition, within a full modernity (Gheorghiu 2008).

This requires also the proposal of a strategy of the valorisation of the place's identity itself, and this can be done through changing individuals' or community's mentality occurring in the area of the basic notion of living.

The intention to dissipate the "heart of the settlement" considered to be the archaeological site through conservation and development policies. In other words, it is intended to guide the tourists to know and go through the village using a route, relying, among others, on the close relationship between the site and the inhabited area (where the exit is).

If it is mentioned, but the inclusion of the tradition in the promotion of cultural landscape, without complexing and exaggerating (as it is about the continuity of a process—tradition—in the context and beyond any era—especially the modern era), this would mean either adapting the old assemblies to the actual needs or acquiring and using some traditional elements (of the stable ones, perennial, convenient) to define the contemporary landscape.

A typological observation that could help this depends on a special feature of the building. If a building has withstood time, this was due to either their viability or the lack of replacement. It was not preserved for cultural or artistic reasons, as we would be tempted to believe. These findings apply to all material components, from home-household to the whole settlement (Fig. 6).

The re-conversion of some buildings in craft workshops (souvenirs), cafés, wine cellar or exhibition, would rhyme the course and interest to redefine the identity and preserving the memory of the place. This place would be managed by the owner himself or rented on short or long term.

Fig. 6 Sarmizegetusa Ulpia Traiana. Almost 60 % of the fences are actually stone walls of the annex buildings, which seem to have lost today, their former function. Minor interventions such as moving the gate beyond the annexes and allowing the access towards the kitchen trough the side wall, would make an urban socket Ochiniciuc (2006), pretext for the access in the building (*source* author)

Navigating the village along the inhabited space of the village tries to steer the tourist towards the common areas that include both private and public spaces. It is desired to combine these spaces, using the means of space reconfiguring, the insertion of new features, the spatial change, the insertion of new features, changing the public image and a strategy for sustainable development. In addition, transforming the old spaces by using a new spatial and temporal expression represents a cultural and identity evolution. These spaces will not be opened or radically changed. On the contrary, the interventions will be local and punctual, giving birth to new spaces and equipments of new points of interest related to high social attractiveness. The choice of materials will also contribute to the creation of a place of history, science, action and contemplation.

7 Process: Intervention—Adapting the Landscape

The result will consist of various landscapes, blended in the new configuration of the place, a new fragmentation. As individual items, these places would not tell us much, perhaps, but building places and landscapes allows us to live together the common and unique dimensions of the space, those of anticipation and unforeseeable, controlled or not.

These landscapes, controlled at human scale, tend to bring along a long gone image of the village, re-functioning the spaces and finding a lost identity. If we take into account the fact that the landscape is a representation and that the representation is related to the choice of an image and the acceptance of its vocation to imply itself in these representations, the process becomes the one of reading and listening to the landscape starting from the existing configurations and yet in constant motion.

The latter can take shape and can give a special importance to the common spaces and this is because the place's images result from the cognitive experience and the spatial realities that are part of the individual's and his

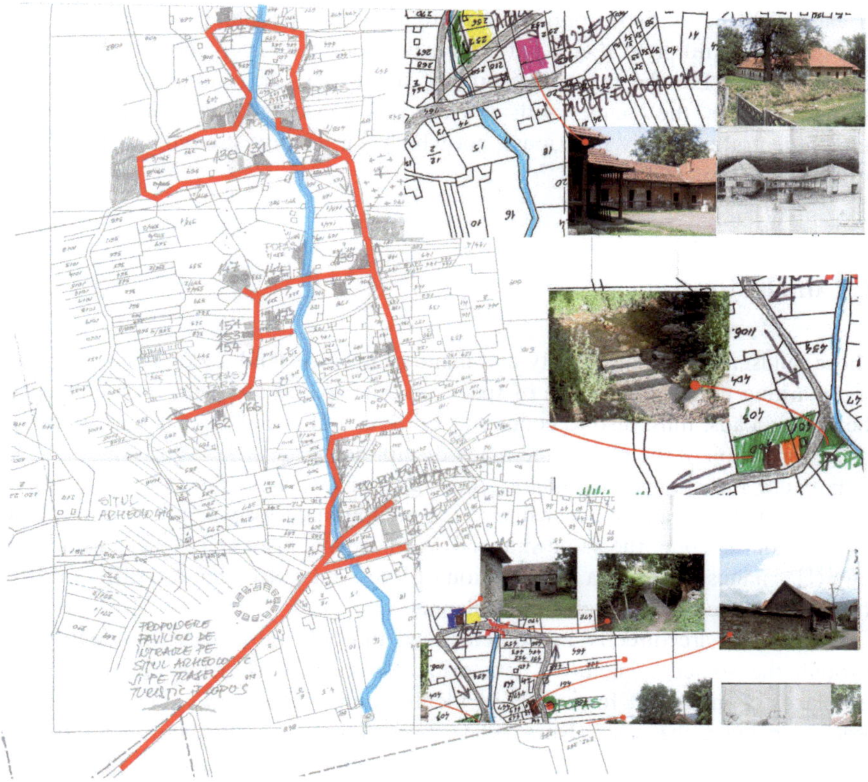

Fig. 7 Route Sarmizegetusa (*source* author and Ochiniciuc 2006) Part of the proposal for a cultural route and remodelling rural that cross the village and reveals its local value. The proposal relates to architectural, environmental and compatibility, as well as the complexity of implementing a social sustainability

community's life. The landscape is defined, in the minds of Jackson (1984), as "a composition of spaces created or modified by man to provide infrastructure or (the background) of our collective existence". The cell spaces, along the path, (different types of landscape) are reserved for operations that are part of a participatory discourse, according to their form and purpose, of a language because "the landscape is a place of conflict, the place of a permanent compromise between what it is established by the authority and dynamics of the vernacular" (Jackson 1984).

These operations of various natures succeed, inform and change mutually. If today they are only minimal interventions, their effects can be huge for the village tomorrow creating a collective memory of the place. Even in the case of a shift of interests, it is modified the hierarchy of different parts of the course and it will be possible to specify the successive choices of interests that accompany it. The revitalization of these areas, apparently lost, assumes an active role in the approach of valorising the touristic approach. The tourism puts people in contact with areas that are not familiar to them, but which provides for acceptance, inclusion and understanding. For the tourists, people of the society, accustomed with the idea of the daily life, the contact with new areas is a visceral approach that does not remain neutral or purely functional, but which, for lecture, strengthens the identity of the place (Fig. 7).

8 Conclusions

These landscapes full of identity and memory become landmarks, places of memory, marking the fragmentation of space, and the spatial reorganization concept of the places leads, thus, also to economic, cultural and social aspects that contribute to the formation of the places and many disciplines.

Because the change is possible, there will always be a ratio of arrangements and combinations ranked according to the predominant form of one of the characters, without keeping the same shape. The landscape, also, can be seen as a living map (Hărmănescu 2011), a composition of lines and spaces not too different from the one produced by the architect or planner.

Differences are directly instrumentalized through the process of identity and through the society's memory: territories, "portions of territory are constituted as 'places of memory', for collective identity and as landmarks for the society in order to manage the past" (Nora 1984). In this sense, the distinctions are usually made between the ordinary landscape, which would be unconsciously produced by a human community and the landscape because the objective is to organize a space that can meet human needs.

Acknowledgments This paper is based on the research work developed within the author during her M.Arch (*Sustainable development—an integrated concept of built space and energetic audit)* and PhD (*Potential And Value Of The Rural Landscape As Part Of The Cultural Landscape. Theoretical approach at a territorial level*), UAUIM.

Bibliography

Choay F (1998) Alegoria patrimoniului. Simetria, Bucharest

Conea I (1935) Clopotiva, un sat din Haţeg—Monografie sociologică. Institutul de Studii Sociale al României, Bucureşti

Creţu VT (1988) Existenţa ca întemeiere. Editura Facla, Timişoara

Farsani NT et al (2012) Geotourism and geoparks as gateways to socio-cultural sustainability in Qeshm Rural Areas, Iran. Asia Pac J Tourism Res 17

Gheorghiu TO (2008) Locuirea traditională rurală din zona Banat—Crişana [Elemente de istorie si morfologie, protective si integrare] Eurobit, Timişoara (in Romanian: Traditional rural dwelling near Banat—Crişana [Elements of History and Morphology, Protection and Integration])

Gusti G (1974) Forme noi de aşezare. Editura Tehnică, Bucureşti

Halbwachs M (1997/1950) La mémoire collective, Les Presses universitaires de France, Paris: Albin Michel
Hărmănescu M (2011) The Landscape—a made and inhabited territory (Romanian: Peisajul—un teritoriu locuit si facut). Peisaj- Arhitectura- Tehnologie- Ambient Ed. Universitara Ion Mincu, Bucuresti, pp 111–118
Jackson JB (1984), Discovering the vernacular landscape, Yale University Press
Lefebvre H (1974) La production de l'espace. Anthropos, Paris
Machedon F (2006) Metoda de analiza morfologica a tesuturilor urbane, Ion Mincu University Publishing, Bucharest (in Romanian: Method of morphological analysis of the urban tissues)
Noberg Schulz C (1986) Genius loci: toward a phenomenology of architecture. Pierre Mardaga
Nora P (1984–1992) Les lieux de mémoire, vol 1. Gallimard, Paris, p 25
Ochiniciuc C (2006) Arhitectura şi schimbarea climatică (Ed), Ion Mincu, Bucharest, pp 98–103
Plesu A (1992) Pitoresc şi melancolie: analiză a sentimentului naturii in cultura europeană (Ed). Humanitas, Bucharest
Popa N (1999) Ţara Haţegului: Potenţial de dezvoltare al aşezărilor omeneşti, Editura. Brumar, Timişoara
Schama S (1999) Le paysage et la mémoire. Seuil, Paris
Thébaud P (2001) Le paysage en quatre dimensions: pour une nouvelle approuche de la concertation, Edition de Kerlan, Coll. "Urbanisme et Paysage", Geo-Vision Avenir
Tudora I (2006) Peisajul/peisajul urban: scurtă incursiune sau o încercare de definiţie, CNCSIS Grant: Spaţiul public şi reinserţia socială a proiectului artistic şi arhitecturalartă, comunităţi urbane, mobilizare, Consorţiu Nr.23/2006, Ed, "Ion Mincu", Bucharest (in Romanian: landscape/urban landscape—sneak peek or attempted definition)
*United Nations conference on environment and development (1992), Rio de Janeiro
Vert C (2002) Tipuri de peisaje rurale în Banat, II-a edn. Editura Mirton, Timişoara

Decoding Agricultural Landscape

Andreea Popa

Abstract Economic and social processes' characteristic of the end of the twentieth century manifested themselves through new ways of organizing space and new patterns of mobility and location of population, activities and labor force, both at urban and regional levels (in the territory adjacent to urban centers). In addition to these changes, extensive urban development, due to demand of land for development and decline of interest in practice primary activities, is reflected in the extra-urban territory, also on landscape perception and use.

Keywords Agricultural landscape • Transformation • Perception • Code • Dynamics • Constitutive elements • Fragmentation

1 Introduction

Extended definitions of the landscape concept constitute an opportunity to consider new territorial landscape types, perceived most often as common productive landscapes, and without being subject of spatial planning policies due to their cultural values. Consequently, these territorial landscapes (rural, cultural, and agricultural) are subject of deep transformation of their features and symbolic, recent concerns looking to preserve their cultural and socioeconomic values, attested by their existence. Neglecting today, the cultural landscapes prevents the modern society from comprehending the importance of the identity of places and the cultural meaning that the places carry with them. It impends on seeing and learning the narrative process of urban perception and evolution that they express during the centuries.

A. Popa (✉)
Department of Urban and Landscape Design, Ion Mincu University of Architecture and Urbanism, Str. Academiei nr. 18-20 sector 1, 010014 Bucharest, Romania
e-mail: apopa.uauim@gmail.com

C. Crăciun and M. Bostenaru Dan (eds.), *Planning and Designing Sustainable and Resilient Landscapes*, Springer Geography, DOI: 10.1007/978-94-017-8536-5_3,

Various methods and techniques are used internationally to assess landscape evolution as territorial system (becoming-assessment-forecasting) which envisages different components of it, related to the scientific approach: geographical, historical, cultural, ecological, etc. The main information/digital tools designed for landscape dynamic assessment are photo interpretation, aggregation of different sets of indicators, digital texts collections analysis. Usually, these sources used in landscape evaluation also contain nonquantifiable sensorial features related to landscape perception (sounds, smells, colors, textures, etc.).

From this perspective, the article presents the main constitutive components of the agricultural landscape as extended territorial system, considered as structural elements and directly influenced by contextual transformations, as well the potential to include them within spatial development policies. In that case, constitutive elements of a landscape are considered formal, functional, morphological, compositional, structural, and inter-related in a particular form, characteristic for each landscape.

2 Agricultural Landscape Concept

Modern definitions of the landscape refer to its cultural meaning: The landscape is the synthesis of space used as a collective space (Jackson 1984) or—as an extension of this definition—the space perceived by the population to whom it gives meaning and shape in accordance to their aims and objectives (Kizos and Spilanis 2004).

By extrapolating these definitions to the landscape practice, it results that agricultural landscape is directly related to the land use, and its approach should include two essential elements: the environmental dimension and symbolic dimension (Kizos and Spilanis 2004) of agricultural landscape.

It is essential for the planners to treat the agricultural landscape, both as a main component of the countryside as rural and cultural landscape system, and as a result of the natural interaction with socio-economic systems. From this perspective, it becomes essential to approach agricultural landscape through its collective representations and significances. Agricultural landscape, through its senses, uses, and significances represents a productive landscape, its existence being primarily defined by the productive function of the territory.

The agricultural landscape should be considered as a principal component of the rural landscape, as a microsystem of it, without excluding the constitutive elements defining the agricultural landscape, which are considered structures of the rural landscape as well. The agricultural landscape is also the result of the interaction between the natural environment and the socioeconomic systems (political factors, location factors related to urban development, cultural factors related to perception, uses and preservation demands for this type of landscape, social factors related to use and organization of this space).

Different evolutions of territory, perceptions and associated practices (Sârbu 2011) are imposing differentiation of agricultural landscapes through various degrees of concentration of cultural/ethnographic elements. Consequently, those who have experienced a slow transformation over time, less influenced by external factors, can be characterized in present as ethnographic testimonies with cultural value, individualized within common agricultural landscape. Maintaining immaterial testimonies of a transformed agricultural landscape or of a lost productive function can indicate the existence at perceptive level of an immaterial agricultural landscape of practices and associated traditions. For example, in former rural areas with an accelerated industrial development and modified profile from rural to urban (professional evolution, activities, education, etc.), customs, traditions, and practices associated to agricultural landscape are still maintained within urban space: Paparudele, Caloianul, Cununa Grâului, Mărţişorul, Babele, etc.

By extending the term "landscape" to the result of interference with the human activities and based on the linguistic origin of the term, the strategic approach of the landscape is considered either as a result of evolution and interaction with human environment or as a cultural symbolic representation.

In accordance with these definitions, agricultural landscape can be defined as surface (Jackson 1984), i.e., space for productive activity and territorial modeling in accordance with the needs of the residing population, and as program, i.e., the sum of the collective significances continuously transformed in accordance with socio-cultural evolution.

> Territory as resource offer the image of a system of which transformation it is inextricable related to place history. Natural and anthropogenic elements (…), clotted in a territorial dynamic open system (…). The territorial and spatial expression of these elements, complex components of cultural landscape are subject of rapidly evolution phenomenon, (…) (Sârbu 2011)

The research proposes to approach of the landscape as an independent and a complex territorial system, related to other urban and rural landscape systems. The premise is that agricultural landscape at territorial level constitutes a complex ecospatial, social, and economical system, more than just a component of other mentioned systems, usually associated strictly to land use or productive functions of the territory. In that case, the agricultural landscape can be differentiated morphologically and significantly form urban–rural landscape; different from the rural landscape, agricultural landscape is conditioned by the productive valuation of territory.

3 Transformation of Agricultural Landscape

The agricultural landscape is continuously transforming as a result and a dynamic expression of the interactions between natural and the anthropogenic interventions: land use and productive activities. Currently, transformations known attributed to the agricultural landscape are considered negative, generating loss of identity and coherence.

Extensive urban development in recent decades caused by the demand for development land and the decreased interest in primary activities are impacting on the surrounding territory. Extraurban development is modifying the landscape, the agricultural, rural or cultural, and is generating new typologies of landscape modeled by urban–rural interactions.

In absence of urban and territorial policies, the agricultural landscape is suffering irreversible transformations, fragmentation, segregations, and alienation, in the end losing its specificity and uniqueness. Any decision at urban level with different spatial and time results is generating new growth models. In that context, diverse approaches of urban development are currently modifying the territorial structure and the way in which the territory responds visually and dynamically to external factors.

The main factors influencing and determining the dynamics of the agricultural landscape are "accessibility" (Antrop 2005), "urbanization" (Antrop 2005), "globalization" (Antrop 2005), "natural disasters" (Antrop 2005), anthropogenic disasters, sociocultural, political-administrative evolutions, and the ownership status.

These factors can be defined in accordance with the particular approach taken, with various implications in terms of severity, extension, and effect. Punctual interventions (at microlandscape level) can totally and irreversibly modify local features, and major interventions can leave unaltered local features (singular elements in landscape: trees, fences, etc.). In the same way, factors which at large scale are not perceptible, such as communication corridors, at the local level can affect and transform the agricultural landscape system: land fragmentation, incompatibilities, structural changes, etc.

4 Constitutive Elements of the Agricultural Landscape

Constitutive elements of the agricultural landscape are the ones composing and contributing to its organization through specific relations which are essential for its structure. The agricultural landscape is defined by its productive features: productive activity and land use, degree of anthropogenic elements.

Territorial scale is one of the most important aspects that defines the constitutive elements of the agricultural landscape. Elements which at territorial scale can be considered structural, at small scale are units: processes and affected surface; elements which at microlandscape level are structural, at territorial level are just punctual elements.

By extending these concepts, we can distinguish the constitutive elements which are most representative for the aggregation of landscape units, based on homogeneity features (Table 1).

The Table 1 outlines the principal constitutive elements identified in case of the agricultural landscape based on the identity and specificity conferred. The classification does not differentiate elements by formal, physical, and functional approaches. This complex approach of the landscape considers that these elements are overlapped: "Determination and analysis of components apart from constitutive elements and different spatial, psychological, economic, and ecological

Fig. 1 Agricultural Landscape Unit—tulips fields in Netherlands

Table 1 Constitutive elements of the agricultural landscape

Constitutive element	
The agricultural landscape unit is represented by homogeneous elements, expanded in the territory: arable land, orchards, vineyards, etc.	Figure 1 Figure 2
The interruption of the agricultural landscape unit is usually a volumetric element; It can become a structural element of landscape through spatial configuration and territorial extension	Figure 3
Edge–corridor–boundary vector elements, i.e., anthropogenic or natural: rivers, communication systems, and technical infrastructures	Figure 4 Figure 5
The landmark is a singular identity in territory. If repeated, it can become a structural element.	Figure 6
The boundary–edge is the characteristic for urban sprawl in territory	Figure 7

characteristics can not control the ensemble" (Bertrand 1978). From this perspective, the inventory of constitutive elements of the landscape is limited to their role inside the landscape unit according to the definitions of (Forman and Gordon 1986) and Lynch (1960). He distinguished five spatial elements based on their evocation capacity for individuals, which can be assimilated to the ecological approach of landscape: district (zone), patch (functional unit), edge–boundary, path–corridor, node–node in matrix of functional units.

Fig. 2 Vineyard Landscape Unit—Tuscany

Fig. 3 Pastures alternating with forestry and shrub areas: interruptions of Landscape Units, Apuseni Mountains

Fig. 4 Anthropogenic corridor and edge: exploitation path, Dobrogea

Fig. 5 Natural corridor: the Danube River, seimeni

Fig. 6 Singular presence in agricultural landscape: a shrub

Fig. 7 Urban–rural edge

Approaching the agricultural landscape dynamics should take into account also the transformations imposed by the sociocultural and territorial context and the relations created between various constitutive elements. To these facts, one can also add sensorial elements which compose the agricultural landscape and offer specificity and uniqueness, through differentiation of the ambiance. Usually, these elements are not used in assessment methods of the landscape change and dynamics and are not included in the preservation policies dedicated to landscape, i.e., colors: tulip fields in Netherlands, lavender fields in France, orchards; smell: aromatic crops; texture and luminosity: olive orchards, vineyards; sound associated with natural elements within the productive landscape, etc.

5 Conclusions

Although the European agricultural landscape protection and enhancement concerns are not recent, concern in Romania for this type of landscape is almost inexistent. The paper brings into question the agricultural landscape with all issues arising as a consequence: definition and conceptual delimitation, protection and enhancement measures. From this perspective, the paper starts from the premise that the agricultural landscape is an essential part of the landscape and a new field of study concerning the specificity and identity of the territories.

Understanding the importance, evolution and potential of the agricultural landscape at local level becomes essential for the local systems to react and to adapt to socioeconomic transformations.

The sector approach of the agricultural landscape generates a fragmented assessment of its problematic, and rarely, its dynamic is understood as synergetic result of all influence factors. The constitutive elements of the landscape are usually treated in a fragmented way, and it is necessary to preserve and valuate them from a complex perspective, starting with the premise that even an ordinary productive territory is a complex independent system with multiple values.

References

Antrop M (2005) Why Landscapes of the past are important for the future. Landscape Urban Plan 70(1-2):21–34

Bertrand G (1978) Le paysage entre la Nature et la Societe, Revue geographique des Pyrrenees et du SudOuest, 49, In: Roger A (ed) (1995) La theorie du Paysage in France, Seyseel, Champ Vallon

Forman RT, Gordon M (1986) Landscape ecology. Wiley, New York

Jackson JB (1984) Discovering the vernacular landscape. Yale University Press, New Haven

Kizos T, Spilanis I (2004) The transformation of landscape: modeling policy and social impacts on the agricultural landscape of Lesvos. Nat Resour Model 17(4):321–358

Lynch K (1960) The Image of the city. Massachusetts Institute of Technology

Sârbu C (2011) Peisaj cultural şi dezvoltare, substanţă şi expresie a dinamicii teritoriului, în Sârbu C *Peisaj Cultural şi Dezvoltare*, Bucureş ti, Ed. Universitară Ion Mincu

Part III
Cultural Landscape: Image/Photo/Space and Time—Memory

Inventing Scotland: Photography, Landscape, and National Identity

Stephen Monteiro

Abstract This chapter considers architectural and topographical photography's role in the formation and expression of national identity in Great Britain in the mid-nineteenth century. It examines opposing depictions of national imaginaries in the 1840s by the English inventor of negative-positive calotype photography, William Henry Fox Talbot, and several Scottish practitioners of that process, including John and Robert Adamson, David Octavius Hill, and the amateur photographers of the Calotype Club of Edinburgh. Talbot's book, Sun Pictures in Scotland, exercised an English claim to photography by following the visual rhetoric of the picturesque in reducing Scottish identity to the ruins and landscapes associated with the work of Sir Walter Scott. The Adamsons, Hill, and the members of the Calotype Club embarked on photography projects that eschewed such British interpretations of Scottish culture, focusing instead on contemporary figures and events, historical sites, and sustained land use. This chapter concludes that although early Scottish photography promoted a cohesive national identity, it set in motion the paradoxical potential for signs of Scottish national character and action to be absorbed into a larger, hegemonic narrative of an Anglicized Great Britain.

Keywords Photography • Calotype • Picturesque • Highlandism • England • Scotland

1 Introduction

In the summer of 1845, William Henry Fox Talbot received a distressed letter from his mother. Five years after the Englishman had invented negative–positive photography and called it "calotypy," Elisabeth Feilding warned her son that his

S. Monteiro (✉)
The American University of Paris, 31 avenue Bosquet, 75007 Paris, France
e-mail: smonteiro@aup.edu

C. Crăciun and M. Bostenaru Dan (eds.), *Planning and Designing Sustainable and Resilient Landscapes*, Springer Geography, DOI: 10.1007/978-94-017-8536-5_4,

Fig. 1 William Henry Fox Talbot, Loch Katrine, 1844, salt paper print from calotype negative, 18.6 × 22.6 cm (7 5/16 × 8 7/8 in), Princeton University Art Museum, Museum purchase, Surdna fund, in honor of Peter C. Bunnell, © Princeton University Art Museum

proprietary claims to the process were in jeopardy. "I have met with a person who told me he had seen in Scotland (photographs) much better than your Talbotypes, that they were called *Calotypes* and were very superior and invented by a Scotchman, a friend of Sir D(avid) Brewster," Feilding conveyed. "(A)ll comes of not having called them Talbotypes *at first*," she admonished him (Schaaf).

Ironically, this confusion came as Talbot published *Sun Pictures in Scotland*, an album of photographs he had taken in Scotland the previous fall. *Sun Pictures* was his second photographically illustrated book, after *The Pencil of Nature*, and the third such publication in the world. Offered by subscription, it comprised 23 salt paper prints from paper negatives that primarily depicted "scenes connected with the life and writings of Sir Walter Scott," according to a publication announcement (Buckland 89). Seven images—or nearly a third of the book—were devoted to Loch Katrine in the Perthshire Highlands (Fig. 1), the setting of Scott's celebrated 1810 poem, *The Lady of the Lake*, and a popular destination for English tourists, writers, and artists, including J. M. W. Turner and John Ruskin (Smith 1989, 27). Unlike *The Pencil of Nature*, whose lengthy captions alongside images representing a variety of pictorial genres emphasized the physical properties and potential applications of calotypy, *Sun Pictures* was a nearly textless endeavor meant to demonstrate the photograph's narrativity. It offered views "impressed by the agency of Light alone" (Talbot 1845, unpaginated) of bucolic sites already widely "impressed" upon the British imaginary through Scott's romantic writings.

Lost in the long shadow of the more ambitious and influential *Pencil of Nature*, *Sun Pictures* has received little attention in the history of photography. It is considered a minor project primarily meant to exploit the popularity of tourism and Scott for the promotion of Talbot's process (Smith 1989, 51; Knazook 128). Studied within the larger context of Scottish photographic production of the era, *Sun Pictures* can be understood instead as an attempt to situate both Scotland and the photograph within the scope of English political and technological hegemony by resorting to British mythologizing of Scotland and its topography as a land whose primary function was to enchant the touring English eye caught in the grip of historical fantasy. In contrast to the Scottish projects considered here, however, Talbot's attempt would fail, lost amid works of the sort referenced by Feilding. When Elizabeth Eastlake would describe the legacy of calotypy in her 1857 account of photography, she would resort to topographical terms to reflect this aspect, noting that "photography made but slow way in England; and the first knowledge to many even of her existence came back to us from across the (Scottish) Border." Eastlake would clarify that "it was in Edinburgh where the first earnest, professional practice of the art began, and the calotypes of Messrs. Hill and Adamson remain to this day the most picturesque specimens of the new discovery" (Eastlake 452).

Talbot's decision to call his invention calotypy rather than "Talbotypy"—in contrast to French rival Louis-Jacques-Mandé Daguerre's decision to name his metal-based, direct-positive process "daguerreotypy"—may have contributed to the misunderstandings voiced by Feilding and implied by Eastlake. But the confusion also stemmed from the achievements of several Scottish photographers, among them David Octavius Hill and Robert Adamson. Foremost among those exploring the technological properties and cultural applications of calotype photography in the 1840s were several artists, scientists, and learned professionals living in and around Edinburgh and the nearby university town of St. Andrews. Esteemed physicist David Brewster, a friend of Talbot's since the early 1830s and an administrator at St. Andrews, had promoted photographic experimentation there through his involvement with the St. Andrews Literary and Philosophical Society and what would come to be known as the Calotype Club of Edinburgh. With lectures, exhibitions, and the circulation of experimental photographs between St. Andrews and Talbot's estate in England, Brewster encouraged his colleagues to shape the material and cultural characteristics of photography.

In an essay on nineteenth-century landscape photography, Jens Jäger observes that "Most modern nations were 'made' in the nineteenth century, and their making was supported by the popularization and redefinition of all kinds of images and the nationalistic (re)interpretation of landscapes and architectural and honorific monuments" (117). Coming at an important juncture in the development of mass imagery and national cultures, these photographic projects made visible competing desires for photography as a means of imbricating cultural ideologies and natural environment in the struggle for political hegemony and national identity. (Indeed, the right to make photographs was part of this struggle. In 1841, Talbot patented the calotypy in England and Wales but—following Brewster's advice—he had not taken out a patent for Scotland, creating an imbalance that heavily favored Scottish development of the process.) In the context of 1845, the proprietary questions raised by Talbot's mother surrounding the photograph extended beyond the simple matter of proper attribution to the wider question of national sovereignty and cultural identity as these emerged around and within the image. Laying attribution for the process at the foot of a Scotsman presented a double affront: credit not only slipped Talbot, but also eluded England in an intrakingdom rivalry.

Fig. 2 John or Robert Adamson, St. Andrews Castle from the east, 1843 or 1844, 12.4 × 17.2 cm (4 7/8 × 6 ¾ in), The J. Paul Getty Museum, image courtesy of the Getty's Open Content Program

As a pictorial category tied to ideas of property and sovereignty, landscape was at the center of this competition. It was a popular genre that lent itself to early photography's need for abundant sunlight and long exposure times. Yet it also offered possibilities for visual expression that negotiated the preexisting discursive terms of painting and printing to permit new relationships between the physical world and the viewer. In particular, the presumed evidentiary value of photographs as images produced by nature itself, when combined with their introduction into the specific cultural and political instability of early Victorian Britain, allowed for a powerful coupling of visual representation and ideology that would nevertheless seem "transparent" and self-evident.

Through its presumed objectivity, the topographical photograph encompassing architectural views, urban scenes, and landscape could be used to further political aims, particularly claims on territory and material culture as markers of national identity, imperial might, or colonial expansion. Working with such notions of the power of photography to preserve or acquire, one can consider the production and performance of photography in Scotland in the middle of the nineteenth century as alternately an act of national resistance and acquiescence to the project of empire. If an English photographer such as Talbot reinforced the colonialist myth of the Other by recording the Scottish landscape and its historical sites as the stuff of literary fancy and a stage for photographic tourism, early Scottish photographers set about building visual narratives of a nationalist sentiment that worked through the vestiges of a national history or character. The focus of Scottish photographers on elements of Scottish political and cultural history, such as the Reformation or the contemporary Disruption of the Church of Scotland and founding of the Free Church of Scotland, demonstrated an interest in recording the traces of a Scottish past tied to its present that ran counter to the concurrent promotion of mythical Highland culture.

In their approach to the landscape and built environments, the Adamsons, Hill, and Calotype Club members presented a more complex vision than Talbot of Scotland and Scottish cultural identity (Fig. 2). While photographs of a bust of Scott and Edinburgh's Scott Monument appeared in the albums of the Calotype

Club, for example, these were presented among images of the ruins of St. Andrews Cathedral and the first Free Church building, Fairlie Free Kirk in Ayrshire, in addition to numerous views of club members' families and homes. All differentiated themselves from Talbot and his underlying reductive aesthetic of the picturesque, promulgated in England by William Gilpin in the late eighteenth century as a means of imagining and representing the countryside (Knazook). They produced a significant number of views that combined topography, architectural elements, and present-day activity with occasional textual references underscoring the lingering impact of a contentious historical relationship with England. These were sometimes meant for publication, as in the case of Hill and Adamson's several book projects around Scottish sites, customs, and industries, but they also took the form of groupings in private albums intended for much more limited audiences. This was the case for the work of Adamson's brother, John, and that of James Francis Montgomery, Hugh Lyon Tennent, and the half-dozen or so other members of the Calotype Club. The small album of views of St. Andrews now known as the Tartan Album, to be considered here in contrast to *Sun Pictures in Scotland*, remains a pivotal document in this story. Created by the Adamson brothers in 1842, Brewster sent the album to Talbot as the epitome of Scottish amateur photography at the moment Robert Adamson decided to open the world's first commercial calotype studio in Edinburgh. With its narrative of resistance and endurance in the face of cross-border hostilities, Talbot may have had the Tartan Album in mind when he travelled to Scotland in the fall of 1844 to make photographs for his own volume.

2 Sun Pictures in Scotland

Sun Pictures in Scotland was published in July 1845, after the fourth of *The Pencil of Nature*'s six fascicles. An illustrated treatise published between 1844 and 1846, *The Pencil of Nature* traced the origins and uses of Talbot's invention by juxtaposing text and image around several subjects and image genres. Significantly, the project was an effort to lay proprietary claim to photography by demonstrating the utility of Talbot's patented invention through reference to personal fortune and historical monuments in documenting Talbot's family home, Lacock Abbey. Through images and text, Talbot described the estate and its history, endowing his images with a proprietary knowledge. With plate 15, *Lacock Abbey in Wiltshire*, Talbot introduced "a series of views representing the Author's country seat," describing it as "a religious structure of great antiquity, erected early in the thirteenth century, many parts of which are still remaining in excellent preservation." Indeed, he closed that description by adding "In my first account of 'The Art of Photogenic Drawing,' read to the Royal Society in January, 1839, I mentioned this building as being the first 'that was ever yet known to have drawn its own picture'," an observation that fused claims of ownership and the self-evidence seemingly inherent to the photograph. With plate 16, *Cloisters of Lacock* Abbey, Talbot detailed the site's founding and mentioned the 1835 publication of *The History of Lacock Abbey*.

In plate 19, *The Tower of Lacock Abbey*, he expounded upon a unique copy of the *magna carta* of Henry II, "the only authority from which the text of this Great Charter can be correctly known." In this effort, Talbot invoked documentation, historiography, and authenticity as guiding traits. Yet it was the accompanying texts, not the image themselves, that voiced them. Indeed, the precious copy of the *magna carta* mentioned by Talbot was present only in the text, presumably within the depicted tower "preserv(ing) an invaluable curiosity" (Talbot 1844–1846, unpaginated). Talbot's gaze in *The Pencil of Nature* was proprietary. For the most part, he (and the reader) would see images of *his* goods and *his* estate.

Sun Pictures in Scotland, on the other hand, can be construed as an attempt to reappropriate calotypy by depicting a seemingly apolitical romanticized landscape that reduced Scotland to the most obvious tourist clichés. In essence, it distanced the calotype from Scottish efforts at self-identity by deploying it to reassert Scottish caricatures fixed within the mythology of lost Celtic Highland culture popularized by James Macpherson and eighteenth-century Ossianism and perpetuated into the nineteenth century through Scott's novels and poems. By relying on the landscape as its primary genre, particularly with Highland views around Loch Katrine, Talbot's album reduced Scottish culture to the ersatz ancient tribal culture described by Scott. To an extent, the book rehearsed the terms of colonial photography. By creating and circulating photographs of colonies and colonized, nineteenth-century empires were able to shape visual epistemologies both at home and in subjugated territories, building normative narratives for the exercise of power and the construction of collective identity (Maxwell; Ryan). *Sun Pictures* shifted from the emphasis on personal property and authority found in *The Pencil of Nature* to one assuming national and imperial scope. *Sun Pictures* opened with images of Edinburgh's Heriot's Hospital and the Scott Monument, followed by four views taken at Scott's home, Abbotsford, three of Melrose Abbey, one of Scott's tomb in Dryburgh Abbey, and seven views of Loch Katrine and the surrounding area. Whisking viewers from the tended grounds of Lacock Abbey in *The Pencil of Nature* to the rugged terrain around Loch Katrine in *Sun Pictures*, Talbot transported them from visions of personal property of national interest (the ancient abbey, the *magna carta*) to the extension of national enterprise in the natural environment.

By the time Talbot's book appeared, Scotland had undergone a long period of assimilation by England toward the promotion of Great Britain as a growing global empire. The Union of 1707 had dissolved the Scottish parliament, clearing the path for Scotland's absorption into the kingdom's fold through a continuing Anglicization meant to reflect a common British culture. At the same time, Anglicization developed a complementary regional narrative of a pure, primitive, rugged people descending from the Highlands. The English—Talbot among them—and many Scots represented Scotland and its culture as inextricably rooted in an ancient Highland culture expressly concocted for that purpose. "As Lowland Scotland becomes more and more like England," remarks Peter Womack, "it turns to the Highlands for symbols and beliefs to maximize its difference" (145).

Murdo MacDonald notes that "By the 1860s the main elements of the tartan, heather and hills stereotype of Scotland were in place" (105). Appropriating and adapting elements of Highland culture like the tartan, while ignoring the forced clearance of those regions, allowed representations of a distinctly Scottish spirit within the British Empire, but rooted in natural forces, thereby diverting attention away from the borderlands and Scotland's painfully entwined political past with England. Indeed, with the clearance of the Highlands, Highlanders were recruited for overseas duties in the Empire, thereby deepening Scotland's role in building the British Empire that circumscribed it. The promotion of the Highlands as a purer, more traditional Scottish culture would appeal to people living in the populous Lowlands as well as to the English. "Any vigorous assertion of national identity would… threaten the English relationship on which material progress was seen to depend," explains T. M. Devine, "and so Highlandism answered the emotional need for the maintenance of a distinctive Scottish identity without in any way compromising the union" (244).

While Scott's stories of the destruction of Highland society in the seventeenth and eighteenth centuries through conflict and poverty could be seen as a corollary of contemporaneous Scotland's destruction under Anglicization, Murray G. H. Pittcock contends that Scott was a patriot, but not a nationalist, placing his faith in the Union to solve Scotland's divisive internecine troubles. "Scott's view of Scotland emphasizes the beneficial necessity of change," Pittcock explains. "In doing so, it promotes two ideas: the inadequacy of Scottish patriotism in coping with historical change, and the incompetence of Scots in ruling themselves" (85). In the perpetuation of Highlandism, from the promotion of the tartan as a national symbol to Queen Victoria's purchase of Balmoral Castle in 1848, Scotland became a colorful component of the kingdom. The Highland myth perpetuated in the eighteenth and nineteenth centuries created a convenient before-and-after scenario that paradoxically gave the Scottish nation an irretrievably lost, yet palpably present, past. "The (Highland) region has national associations precisely because it has been made in the minds of outsiders," states Charles Withers. "The historiographical creation of how we have *believed* the Highlands to be has been both more enduring and more fascinating (and enduring because it has been fascinating) than our knowledge of changes in Highland life and economy" (155). *Sun Pictures* is a compelling example of that historiographical creation, made by the mind and eye of a prominent outsider.

3 The Picturesque

In *Landscape and Power*, W. J. T. Mitchell asserts that one should ask "not just what landscape 'is' or 'means,' but what it *does*" in regard to the expression of authority. For him, landscape is a medium, rather than a genre, and closer to a verb than a noun in its function. "Landscape has a double role with respect to something like ideology," Mitchell asserts,

> It naturalizes a cultural and social construction, representing an artificial world as if it were simply given and inevitable, and it also makes that representation operational by interpellating its beholder in some more or less determinate relation to its givenness as

> sight and site. Thus, landscape (whether urban or rural, artificial or natural) always greets us as space, as environment, as that within which 'we' (figured as 'the figures' in the landscape) find—or lose—ourselves (2).

In this operation, the late-eighteenth-century introduction of the picturesque as an aesthetic category in England and its broader application to the topography of the British Isles could be seen, in itself, as part of the colonial enterprise. Early English photographic journals commonly included descriptions of excursions around Britain where amateur photographers could reduce the varied terrain to a series of symbolic images of the "rough but not unpleasant nature" that typified the picturesque (Jäger 123). An amateur photographer writing in *Photographic Notes* in 1858, for example, describes his visit to Melrose, Kelso, and Dryburgh, Scotland (some of the same locales Talbot photographed), "as possessing many and great points of interest, whether to the ordinary traveller in search of the picturesque, or to the primary and more searching eye of the photographer, who seems never satisfied, not even with walking away with Abbeys in his portmanteau, and bridges, rivers, ruins, and palaces, under his arm, he still craves for more" (Gutch 137).

With *Sun Pictures*, Talbot sought to capitalize on the picturesque fascination with the Scottish Highlands produced by the success of Scott's works. As Robin Lenman explains, "Even before 1800, a combination of Ossianmania and Continental war was encouraging travelers to go north. But it was the extraordinary Walter Scott phenomenon in all its manifestations—including George IV's visit to Edinburgh in 1822, stage-managed by Scott—that laid the foundations for a whole industry" (93). By 1844, the trek had become more like a pilgrimage. Turner visited Scotland several times and produced watercolors of the Highlands to illustrate *The Poetical Works of Sir Walter Scott* in 1833. The places Talbot depicts—from the Scott Monument in Edinburgh to Abbotsford, Melrose, Dryburgh, and Loch Katrine—faithfully follow the itineraries suggested by guides like *Black's Picturesque Tourist of Scotland* (Taylor 21). "The idea of illustrating views connected with Scott's writings in itself was hardly an original one," notes Graham Smith, "What was unprecedented was that Talbot's were photographic views, 'the sun-pictures themselves'" (1985, 27). This naturalizing gaze of the camera inscribed the views based on Scott's Highland mythologies within a mythologizing realism, laying claim to these places within an imaginary that constructed a vision of Highland culture as both quintessentially Scottish and a spectacle within a more and more Anglicized Britain. It situated the picturesque as extant in the countryside, emanating neither from the hand of the artist nor the eye of the viewer, thereby "finding" English aesthetic categories embedded in the Scottish wilds (Fig. 3). As Jäger claims, all landscape images "were subjected to the same picturesque gaze. The confirmation of regional identity did not contradict the emergence of the interpretation of these images as 'British.' However, what exactly 'British' meant remained vague…." (128–129).

In its English origin, the picturesque nevertheless acted as an English frame for the image. The photograph not only subjugated the Scottish landscape to this operation of English visuality, but also implied through its "natural" means of production

Fig. 3 William Henry Fox Talbot, the tomb of Sir Walter Scott, in Dryburgh Abbey, 1844, salt paper print from calotype negative, 16.8 × 17.9 cm (6 5/8 × 7 1/16 in), The Metropolitan Museum of Art, the Rubel collection, purchase, Lila Acheson Wallace and Jennifer and Joseph Duke Gifts, 1997, © The Metropolitan Museum of Art

(i.e., the "sun picture") that this colonizing visual category was already present in nature. This imbrication of Scottish landscape and English aesthetics served the contemporaneous conflation of "English" and "British" within the kingdom. As one Scotsman claimed, "Deliberately, systematically, and pertinaciously, (the English) have insisted upon representing England as the Empire, and the Empire as England" (Burns 15–16). In the English "regard" of these landscape photographs, there was nothing vague about the British: the Scottish landscape appeared to have been ensconced within a totalizing English gaze before an Englishman even set foot on a Scottish heath. "In the case of the Scottish Highlands," Stephen Copley and Peter Garside observe, "the combination of political repression, economic exploitation, and aesthetic sentimentalisation of the Scottish landscape in the early nineteenth century clearly renders the Picturesque 'invention' of the region a hegemonic cultural manifestation of the English colonizing presence" (6–7).

In this vein, *Black's Picturesque Tourist of Scotland* quoted liberally from *The Lady of the Lake* and other sources in describing the Scottish landscape, offering travelers the opportunity to match physical reality to literary fantasy. Scott's evocation of Loch Katrine's hills as "earth-borne castles" whose "briar rese fell in banners green" translates natural features into acceptable signs of civilization and culture (194). As a picture book, *Sun Pictures* could be understood as a companion piece, offering visions of the landscapes described in the poems to create a powerful association between text and image, fiction and reality. In this context, Scotland became a liminal entity, neither entirely absent nor present, but a significant piece of the totalizing English imaginary. Because the audience had already internalized the story it illustrated, the album could function without text in inhabiting the silent space between this familiarity and the uncanny presence-and-absence embodied by a photograph.

4 The Tartan Album

Two years before Talbot's photographic tour of Scotland, an album had arrived at Lacock Abbey that deployed the calotype process to very different ends. On October 22, 1842, Brewster had informed Talbot that Robert Adamson was "preparing a little book containing his best works" to be presented to the English inventor. "You will recollect of promising your countenance and support to any person whom I could induce to practise the Calotype as a profession in Scotland," Brewster reminded his friend. "When the specimens reach you, you will be able to judge of Mr. Adamson's merits, and to do what you can to support him" (Schaaf). The small, leather-bound album reached Talbot a month later with a letter from John Adamson—Robert's older brother and his tutor in photography—hoping that the images "will not be devoid of interest from the objects which they picture, whatever may be their rank as specimens of the art" (Schaaf; Harley and Harley 297). The album and its "calotype gems" (to use Brewster's words to Talbot) were the antithesis of the images that would be published in *Sun Pictures*. It is unclear when this gift garnered the misleading title of the Tartan Album, a designation that abundantly reflects the distorted characterization of Scottish culture by Highlandism. Indeed, the album's 18 images make no references to the Highlands, instead illustrating Lowland history and culture through views of historical monuments and sites in and around St. Andrews. Rather than emphasizing picturesque properties (which Eastlake would *a posteriori* attribute to Hill and Adamson's work in 1857), these views allude to episodes of oppression and violence at the root of Scotland's antagonistic relationship with England, reinforced by captions (in John Adamson's hand) that explicitly reference that history.

The album includes four portraits (among them an image of Brewster affixed to the inside cover as though he were the progenitor of the project), a photograph of statuary, and another of figures with a horse. The remaining 12 photographs (all of which immediately follow the Brewster portrait) depict buildings, ruins, churchyards, and farmland in St. Andrews and the Adamson family seat of Burnside. The ruins of Blackfriars Monastery are followed by two views of the ruins of St. Andrews Castle, two views of the ruins of St. Andrews Cathedral, two views of North Street and St. Salvator's College Chapel, one image each of the Episcopal Chapel, West Port, the view from Brewster's garden wall, St. Mary's College, and the Burnside countryside. Unlike the image titles in *Sun Pictures*, which appear only as a list at the front of the book, titles in the Tartan Album appear facing or below images and are often followed by significant historical information difficult to glean from the photograph. Referring to the establishment of the university, the castle, and the church, as well as the violent events of the Reformation, they stake out key components of Scottish cultural identity and national struggle. Through these descriptions, a viewer unfamiliar with Scottish history would learn that the monastery and cathedral had been "demolished by the followers of (Protestant reformer) John Knox," that the castle dates to 1200, St. Salvator's College to 1455, and St. Mary's College to 1538 (Harley and Harley 308–312). Such details

Fig. 4 David Octavius Hill and Robert Adamson or Robert Adamson alone, St. Regulus tower and the east gable of St. Andrews from the northwest, ca. 1845, salt paper print from calotype negative, 14.5 × 19.2 cm (5 11/16 × 7 9/16 in), The J. Paul Getty Museum, image courtesy of the Getty's Open Content Program

diminished both the romantic and picturesque potential of these images. Ruins were no longer the mysterious productions of nature and time, but the physical traces of specific moments of political upheaval. The events mentioned are much more complex than their captions suggest, however. For example, Knox was a Scottish reformist who alternately worked with and against the British crown in the upheaval of sixteenth-century Britain, complicating his role in the development of Scottish nationalism. Nevertheless, these visual and textual cues prevented the seamless insertion of these views into an imaginary of stable British identity. For a view of the ruins of the cathedral and the tower of St. Regulus Chapel, which is nearly identical to a view found in a contemporaneous album kept by Brewster (Fig. 4) and that also appeared in Hill and Adamson's 1846 *Series of Calotype Views of St. Andrews* (Smith 1990, 115; Stevenson 1981, 203), Adamson added that the tower is "commonly called the 'square tower' and is) supposed by many ancient Scotch writers to be as ancient as the 4th or beginning of the 5th century (but is) certainly as old as the middle of the 9th century" (Harley and Harley 310). These details pointed to the vestiges of an ancient, highly developed society that predated the Norman conquest of England by at least two centuries. More important, perhaps, Adamson invoked a long-standing, indigenous historical tradition to rival any British historicization of the Scottish and their land.

In a study of the Tartan Album, Ralph and Joanna Harley emphasize its role in visualizing history by noting its relationship to both nineteenth-century preservationist efforts in St. Andrews and Brewster's 1843 account of photography, written at the time of the album's assembly and published in the *Edinburgh Review* (Harley and Harley 295–296). St. Andrews was the ancient ecclesiastical capital of Scotland and home of its oldest university. According to Brewster, St. Andrews' Literary and Philosophical Society promoted not only calotypy but also the preservation of the area's historic sites. Indeed, the society had fought successfully in 1839 to stop the sale of the cathedral's ancient boundary ruins by the Commissioners of her Majesty's Woods and Forests, an event that pitted Scottish intellectuals against the British government's wishes to profit from the passing into private hands of a site significant to Scottish history. As the society's secretary and curator of its museum, John Adamson was closely involved in such efforts. This link between photography and preservation emerged in Brewster's *Edinburgh Review* article, where he articulated a belief in landscape photography's historiographical potential that perhaps derived from his participation in such events. "The self-delineated landscape is seized at one epoch of time, and embalmed amid all the co-existing events of the social and physical world," Brewster claimed, concluding that "every picture becomes an authentic chapter in the history of the world" (64–65).

5 The Studio of Hill and Adamson

Graham Smith identifies the Tartan Album as a "turning point" between Scottish photographic experimentation based in St. Andrews and its professional exploitation in Edinburgh, where Robert Adamson opened a studio in the shadow of Edinburgh Castle in summer 1843 and shortly thereafter joined into a partnership—with Brewster as matchmaker—with landscape painter David Octavius Hill. The studio would become an Edinburgh cultural institution by the time Talbot published *Sun Pictures*. Considered one of the most important and successful photographic undertakings of the nineteenth century, the partnership sprang from the exigency of Scottish political events. Hill's entrance into photography from painting arose from his desire to document and commemorate the founding of the Free Church of Scotland with a painting commemorating the Disruption—that moment in May 1843 when 400 ministers jointly deserted the Church of Scotland over state intervention in the church and the reallocation of responsibility from church elders to local landowners in appointing ministers. Hill and Adamson's partnership was forged when Hill abandoned sketching the Free Church founders for the convenience of photographing them.

The pair envisioned other photography projects, however, that would transform their collaboration into a major effort to depict Scottish history and culture. In August 1844, they announced their intention to publish six series, entitled *The Fishermen and Women of the Firth of Forth*, *Highland Character and Costume*, *The Architectural Structures of Edinburgh*, *The Architectural Structures of Glasgow &c*, *Old Castles, Abbeys &c in Scotland*, and *Portraits of Distinguished Scotchmen*. Hill had earlier executed dozens of paintings for *The Land of Burns*, an 1840 publication of views related to Robert Burns' poetry, and the subjects of these photographic series clearly indicated the pair's interest in depicting specifically Scottish subjects ranging from the ancient to the contemporary. All based on subscriptions, none of the series was fully realized, nor did any gain widespread notice. Hundreds of negatives were made, however, and their prints appeared in several albums assembled for friends and institutions. Hill and Adamson's work included such pioneering efforts as the documentation of the dwindling fishing community of Newhaven and the endangered monuments and ruins of St. Andrews and Edinburgh—including buildings such as the John Knox house, slated for demolition—as Scotland moved toward greater industrial, economic, and political harmony with England. As their photographs circulated and drew the attention of the British press, Adamson was misidentified as the calotype's inventor. A month after Feilding admonished Talbot regarding such erroneous claims, Hill himself would contact Talbot to explain. "I have written the editor of the *Liverpool Standard* and have put him right both as to the Discoverer and the period of the discovery of the Calotype," he wrote, transmitting the text of that letter. In pointing out the error, Hill had informed the publication that while Adamson did not create the calotype, he "had laboured, with some success, in improving its processes" (Schaaf).

6 Conclusion

A year after the publication of *Sun Pictures in Scotland*, publisher John Murray asked Talbot for examples of his photographs to "facilitate a comparison between the art in Edinburgh & London" to be executed by Elizabeth Rigby, who as Elizabeth Eastlake would publish the 1857 essay on photography cited earlier. "I cannot refrain from wishing that it were possible for Mr. Hill to act in conjunction with you," Murray added. "There are points in which your Calotypes have the decided superiority over his—there are others in which I think he excels—especially in obtaining artistic effects—a combination of the two would be a step in advance" (Schaaf). Murray's words not only reflected the widespread perception at that time that Hill and Adamson's photographs primarily represented the vision of Hill, but also conveyed the sentiment that the Edinburgh studio's work had a visual sophistication lacking in Talbot's efforts. Whatever Murray's wishes, however, not only were Talbot's photographic exploits in Scotland behind him, but by 1846, he had given up making calotypes altogether, turning his attention instead to the development of photogravure processes that might facilitate the publication of photographically illustrated books. With Adamson's untimely death in 1848, the Edinburgh studio of Hill and Adamson would also close its doors, only five years after its founding.

The example of topographical views of Scotland from 1841 to 1845 illuminates the complex, interlocking relationship that developed in mid-nineteenth-century Britain between photography, landscape, ideology, and questions of national identity. Talbot's emphasis on a derivative, bucolic, fantastical vision of the Scottish landscape and culture easily fits within Anglo-British colonialist mythologizing of ancient Highland clan culture embodied in the figure of the noble savage made famous by Scott. Early Scottish photographers, particularly the members of the Calotype Club, Hill, and the Adamsons, produced an alternative vision of Scotland that privileged sites of culture and current events that distanced both photography and the representation of Scotland from an Anglicizing perspective.

This Scotland was not always in line with English universalizing perceptions or depictions. Its photographs referenced specific episodes in Scottish historical and cultural development through portraiture, cityscapes, and architectural and genre studies. Considered jointly, the work of Talbot and early Scottish photographers raises important questions about the role of nationalism within the construction of empire and the ways in which visually centered activities like tourism or social documentation can contribute to its operation. While the 1840s were marked by political unrest and economic depression, the 1850s would be a time of prosperity witnessing the reassertion of British might in the world through victories in the Crimean War and the Indian Rebellion of 1857. By the end of that decade, Scottish photography would fall into rank by perpetuating the tourist trail with images of tourist sites or rustic scenes. With the twin rise in the 1850s of the wet-plate collodion photography process and *carte-de-visite* photographs, landscape views would circulate well beyond the limited range of subscription folios and personal albums. Between 1855 and 1863, Scottish professional photographer George Washington Wilson would increase his sale catalog from 44 Scottish landscape views to over 400 views of landscapes around Britain (Jäger 122). In comparison, *Sun Pictures* had sold only 103 copies in 1845, though its list of subscribers included Queen Victoria (Buckland 89).

Scottish photography would regain a sense of place and history with the public photographic surveys of the second half of the century, including the city of Glasgow's commissioning of Thomas Annan's *Views of the Line of Loch Katrine Water Works* (1859) and *Photographs of Streets, Closes &c. Taken 1868–1871* (1871). Unlike the work of the St. Andrews and Edinburgh photographers of the 1840s, these series documented modernizing public works and urban renewal projects that would leave behind the remnants of the past to serve Scotland's future. Annan and others would depict the disastrous "ruins" of contemporary urban living. The Scottish landscape was changing rapidly, and with the demise of the picturesque, photography would be employed as a convenient and convincing means to record the present as soon-to-be past in the face of Britain's increasingly large-scale industrialization.

References

Black A, Black C (1852) Black's picturesque tourist of Scotland, 9th edn. A. and C. Black, Edinburgh

Brewster D (1843) Photogenic drawing, or drawing by the agency of light. Edinb Rev 104:209–342

Buckland G (1980) Fox Talbot and the invention of photography. D.R. Godine, Boston

Burns W (1861) What's in a name?. Thomas Murray and Son, Glasgow

Copley S, Garside P (1994) Introduction. In: Copley S, Garside P (eds) The politics of the picturesque: literature, landscape, and aesthetics since 1770. Cambridge University Press, Cambridge, pp 1–12

Devine TM (1999) The Scottish nation: a history, 1700–2000. Viking, New York

Eastlake E (1857) Photography. Q Rev 101:442–468

Gutch JWG (1858) Recollections and jottings of a photographic tour, undertaken during the years 1856–7. Photogr Notes: J Birmingham Photogr Soc 52:136–137
Harley RL Jr, Harley JL (1988) The "Tartan album" by John and Robert Adamson. Hist Photogr 12:295–316
Jäger J (2003) Picturing nations: landscape photography and national identity in Britain and Germany in the mid-nineteenth century. In: Schwartz JM, Ryan JR (eds) Picturing place: photography and the geographical imagination. I.B. Tauris, London, pp 117–140
Knazook E (2009) A picturesque photographic tour through Scotland. Genre 29:127–159
Lenman R (2003) British photographers and tourism in the nineteenth century: three case studies. In: Crouch D, Lübbren N (eds) Visual culture and tourism. Berg, Oxford, pp 91–108
MacDonald M (2000) Scottish art. Thames & Hudson, New York
Maxwell A (2000) Colonial photography and exhibitions: representations of the native and the making of European identities. Leicester University Press, Leicester
Mitchell WJT (1994) Introduction. In: Mitchell WJT (ed) Landscape and power. University of Chicago Press, Chicago, pp 1–4
Pittcock MGH (1991) The invention of Scotland: the Stuart myth and the Scottish identity, 1638 to the present. Routledge, London
Ryan JR (1997) Picturing empire: photography and the visualization of the British empire. University of Chicago Press, Chicago
Schaaf LJ (ed) (2003) The correspondence of William Henry Fox Talbot. Database. http://foxtalbot.dmu.ac.uk/index.html. Accessed 10 Oct 2013
Smith G (1989) Sun pictures. University of Michigan Museum of Art, Ann Arbor
Smith G (1990) Disciples of light: photographs in the Brewster album. The J. Paul Getty Museum, Malibu
Stevenson S (1981) David Octavius Hill and Robert Adamson: catalogue of their calotypes taken between 1843 and 1847 in the collection of the Scottish national portrait gallery. National Galleries of Scotland, Edinburgh
Stevenson S (ed) (1995) Light from the dark room: a celebration of Scottish photography, a Scottish-Canadian collaboration. National Galleries of Scotland, Edinburgh
Talbot WHF (1844–1846) The pencil of nature. Longman, Brown, Green and Longmans, London
Talbot WHF (1845) Sun pictures in Scotland. Privately printed, London
Taylor R (2007) Impressed by light: British photographs from paper negatives, 1840–1860. Yale University Press, New Haven
Withers C (1992) The historical creation of the Scottish Highlands. In: Donnachie IL, Whatley CA (eds) The manufacture of Scottish history. Polygon, Edinburgh, pp 143–154
Womack P (1989) Improvement and romance: constructing the myth of the Highlands. Macmillan, Basingstoke

Photography as Exercise of Memory in Order to Reassess, Recompose and Revive Landscapes

Andreea Popa

Abstract The article outlines the main features of landscape photography, as tool for discovering landscape change additionally to cultural and perceptive transformations of the context. The main problems addressed by the article, in relation with photography as interpretation tool, are the temporal evolution of landscape, specificity of place, constitutive elements of landscape, and the way in which photography can modify the perception of a landscape. In accordance, photography can be interpreted as a descriptive image of constitutive (structural) elements of a landscape. A known landscape, without suffering any transformation, can modify its image due to lack of structural representation elements perceived by a viewer. In last years, we face increased interest for using photography as tool to identify and interpret landscape change. We use photography as representation and description of a landscape, as testimony for its evolution, identity and features. From this perspective, decoding a photo can reveal specific elements, features of a landscape or can lead to discover a new landscape, modified by cultural, sociological and perceptive factors; in landscape, study becomes important the way in which photography is used and understood as interpretation and representation tool.

Keywords Photography • Memory • Composition • Assessment • Revival Specificity • Space and time • Reconstruction • Lost landscape • Place

A. Popa (✉)
Faculty of Urban Planning, Department of Urban and Landscape Design, "Ion Mincu" University of Architecture and Urbanism, Str. Academiei nr. 18-20, Sector 1, 010014 Bucharest, Romania
e-mail: apopa.uauim@gmail.com

C. Crăciun and M. Bostenaru Dan (eds.), *Planning and Designing Sustainable and Resilient Landscapes*, Springer Geography, DOI: 10.1007/978-94-017-8536-5_5,

1 Introduction

Today's landscapes (rural, productive, coastal, urban, rural, territorial, etc.) differ greatly from those developed through past times. Until last century, without present progress, development and demand for resources, the use and consumption of landscape (and its components: land, resources) caused a little disturbance to the natural macro-landscape. Due to last century socio-economical and technological transformations, landscapes are deeply influenced and transformed, and in some cases, they evolve or change totally, until disappearance.

In order to identify landscape change, an important technique is photography, used in order to reconstruct past landscapes or to evaluate the magnitude and impact of landscape change (Magill and Twisse 1965).

A daily action, like watching a series of photographs, can induce to the receiver different memories and emotions; all of these are the result of a different perception, of decoding an image based on a different set of elements. Starting from this assertion, the article outlines the main components of landscape that is, or should be represented by photography in order to use it in scientific sense, and interpretation that can be related by decoding a photo by different subjects.

Lot of research themes have tried to discover or outline the relation between photography and landscape, approaching photography as a frame of a certain landscape, as introspection to memories related to a landscape (Pocock 1982), as tool in order to valuate and disseminate cultural heritage (Molyneaux 1997), as scientific method in order to assess landscape transformation (Kull 2005) or as descriptive image of a temporary state of landscape. Common uses of photos to measure cause–effect events in ecosystems range from short-time to long-term time lapse (Gruell 1980). From cultural perspective, the connections or uses of photography are multiple and various, depending on the approach or discipline involved.

The article approach is based on the defining elements of a landscape (discussed in the article as constitutive elements of a certain landscape) and the relation with visual representation (as photography), trying to decode the way in which the photography describes a past landscape, outlines the specific elements of a landscape or induce to the viewer the idea of specificity and identity due to its compositional and visual value (in accordance with semiotics theories).

The photo captures just a frame of the landscape, both dimensional (included with the eyes) and temporal (a specific moment or state). In the article, I opted to use also the terms of space and place, which can describe the frame captured by a photo in relation with specificity and temporality of the landscape.

Thus, the essay is generated by a set of questions regarding the relation between photography and landscape from semiotic approach and tries to decode the specific language and signs which are possible to connect the photo and the viewer by creating a certain (perceived) image of a landscape (related not only to its elements but also to its significance), in order to use photography not only for landscape analysis (evaluation of landscape change) but also to recreate, recompose or revive lost landscapes.

2 Definition of Terms

Landscape can be defined in multiple ways, depending on the discipline. One of the definitions which attest the cultural value of any landscapes refers at a cultural image, a medium of representation and symbolisation of events through images (Daniels and Cosgrove 1994). The uniqueness of the landscape experience for individuals or groups shows that all landscapes are cultural (Cosgrove 1994). Modern definitions of the landscape refer to its cultural sense: the landscape is the synthesis of space, used as a collective space (Jackson 1984), or by extension of this definition, the space perceived by the population to whom it gives meaning and shape in accordance to their aims and objectives (Popa 2012).

By extrapolating these definitions to the practice, results that even the common landscape (the entire territory in accordance with Landscape European Convention) contains cultural features (related to perception, habits, socio-economical evolution and connotations). Landscape concept refers to subjects who perceive and use the landscape in different ways (production, consumption of landscape) and for various purposes (Popa 2012). From this point of view, any approach of the landscape should refer to its symbols for different groups of individuals, and the study outlines the relation between different perception and interpretation of image (photo) in order to extract landscape features.

Georges Bertrand assumed that "determination and analysis of separate components of constitutive elements and diverse spatial, psychological, economical and ecological features cannot control the ensemble". The landscape complexity is morphological (form), constitutional (structure) and functional (activity) and should not reduce it by division" (Bertrand 1978).

"By imposing the spatial organisation specific to a territory (we cannot talk about specificity in an object on its own or part of an urban space), it is not made up on the basis of a unique feature, but a group of general features, centralised during a long period of time under the polishing action of a large multitude of factors (…)" (Sandu 2007). In accordance with that definitions, the article discuss as constitutive elements of a landscape, all its constituents: formal, cultural, spatial, structural, functional, in order to identify how a photo express a landscape, outline its specificity or reproduce a specific landscape. Approach of this article is focused mainly on formal elements (spatial and morphological) but also on related structural or functional elements which can be captured or represented by an image (activities, traditions and atmosphere).

3 Constitutive Elements of Landscape, Coordinates in Landscape Photography

An important element in order to define landscape structure is territorial scale of analysed landscape. In accordance, structural elements, at territorial scale, at micro-scale constitute independent functional units; elements, which constitute

Fig. 1 Agricultural landscape. **a** Multiple functional units, Cluj County, Romania. **b** Multiple functional units, Alba County, Romania. In both *photos*, same constitutive elements (agricultural land, patches, orchards) with different density and disposal

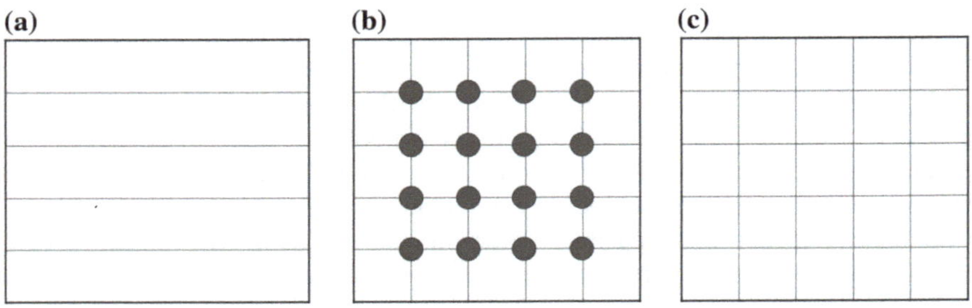

Fig. 2 Landscape diagram, decomposing Fig. 1a, b in constitutive elements. The specificity of site is given by their disposal, overlapping and density

structural elements at micro-scale, are perceived as punctual ones at territorial level, without having a defining role. From this point of view, photo analysis and interpreting should capture and discuss the set of constitutive elements, characteristic for a certain landscape, taking into account the scale.

Related to perception and structure of landscape, main elements distinguished both by perceptive studies (Lynch 1960) and ecological studies (Forman and Gordon 1986) are also outlined by photography methodologies in order to assess landscape change (Magill and Twisse 1965): district (Lynch 1960)—patch (Forman and Gordon 1986), limit (Lynch 1960)—boundary (Forman and Gordon 1986), path (Lynch 1960)—corridor (Forman and Gordon 1986), node (Lynch 1960)—network and matrix (Forman and Gordon 1986), landmark (Lynch 1960). These elements are contained by each captured image of a certain landscape, analysing in temporal series the existence and amplitude of changes indicating landscape transforming. Photos containing same constitutive elements with similar density and disposal can indicate either same typology of landscape, either, due to lack of representative elements, a different landscape, interpreted the same based on decoding of a pre-printed visual language (Figs. 2, 3, 5).

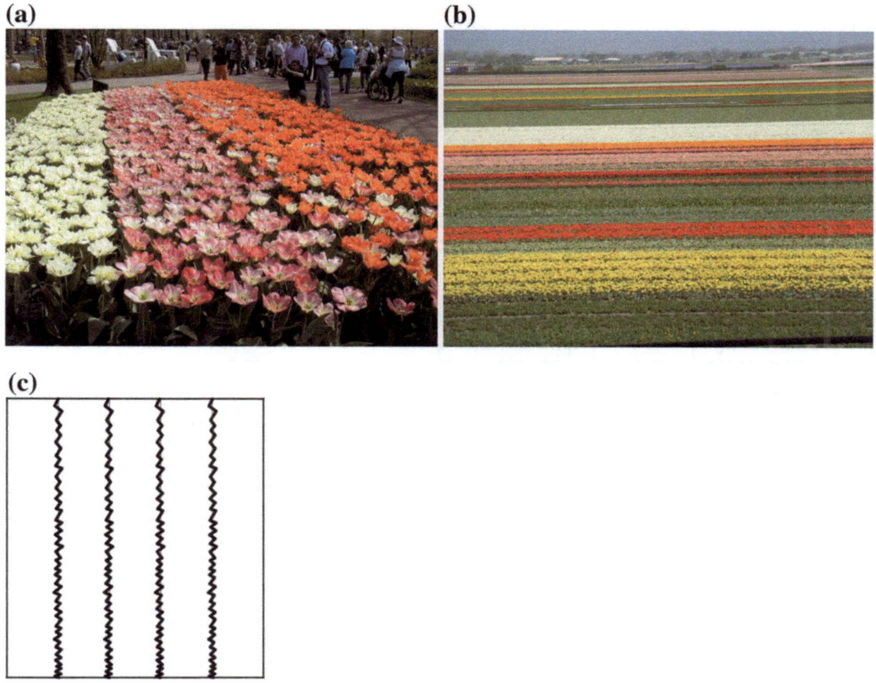

Fig. 3 Tulips landscape. **a** *Rows* in urban park. **b** Agricultural landscape. **c** Landscape diagram, decomposing (**a**, **b**) in constitutive elements. Same perception of different landscapes as result of decoding the scale of the elements

Fig. 4 Urban landscape. **a** Cluj Napoca, Romania. **b**. Vilnius, Lithuania; in both *photos*, same constitutive elements, with different disposal, texture and details

> In first faze of structural semiotics, the writings were about analogy codes through which photos denote objects, connotations codes through which denotation serve to a secondary system of significance and about the "rhetoric" codes of overlapping elements inside a photo and between different photos, but similar (…) Each photo signify on basis of plurality of codes, with various type and number from an image to another (Burgin 1982).

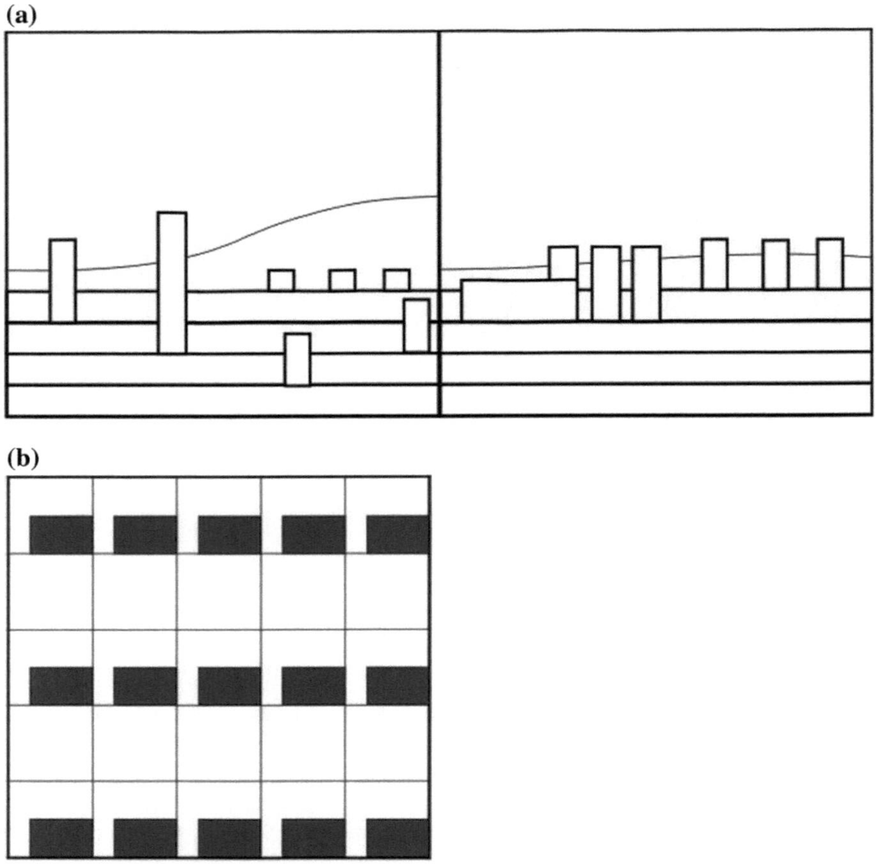

Fig. 5 Landscape diagram, decomposing Fig. 4a, b; Difference in perception is given by texture and disposal and density of landmarks

A photo can represent a modified landscape, which we recognise due to visual codes used by the artist. The compositional elements of the image (constitutive elements of a landscape): patch corridor, node, boundary, landmark and texture represent codes that can be found by the receiver in present space, even if the detailed elements are modified. The receiver recognises the significance of the punctual elements, spatial composition and image sequencing. By combining codes and significances, the photo suggests the linguistic structure of a landscape, by describing the ambiance and place specificity.

Punctual elements (landmarks) that structure vertically and imprint place atmosphere constitute a visual element which if it is removed or disposed different changes the significance of the image. In systemic approach, removal of one element does not destroy the landscape system, just modifies its visual structure; role of the lost element is taken by another existent one. Insertion or removal of an element, grafted on mental image of the place can barely modify it, and the mental perception reconfigures the relations which are the basis of primary decoding that landscape.

> The intelligibility of a photo is not a simple thing; the photos are written texts in terms of what we can name a "photographic speech", but that speech like any other is the place of complex integration, a spatial overlap of a series of anterior texts considered "for granted" in some cultural and historical context (Burgin 1982).

Landscape dynamic can be studied as temporal and spatial reorganisation of the territory due to social and economic demand at a certain evolutionary stage (Popa 2012). Interventions that generate landscape change, differ related to their amplitude (structural or not, intensity, extension, etc.). As result, related to scale of the landscape interventions can be classified and correlated to photo interpretation. Photographic interpretation of landscape change should also identify the amount of changes and their impact on landscape structure, change in activities that generate and evolve in the landscape and elements related to individual perception.

4 Temporal Evolution and Specificity of Landscape Through Photo Decoding

A specific combination of formal (spatial, structural) and non-formal (socio-economical, cultural, traditional and perceptive) factors represents the main feature that confers to a landscape specificity and identity. Landscape identity and specificity are also related to collective perception, meaning the representation and sense for a group of individuals (which use it, living it, lecture it and giving it new senses in accordance with their experience). That differentiation of landscapes, in terms of identity and specificity is outlined by this article: photo is just a representation that contains specific and defining features of a landscape, or it is a tool that in some cases creates specificity in accordance with collective perception?

"An urban existence is defined by a series of factors that, in their turn,—inevitably—undergo a series of modifications. Such modifications convergence may result, either fortuitously or naturally, consciously or not, a partial or even total deterioration of urban existence at a given moment. We may, consequently, have to face either a general replacement of an urban existence or its completion through an end (usually when such an urban existence further continues to offer a functional, economical or spiritual interest)" (Sandu 2007). These evolutionary transformations represent the essence of landscape which is evolving due to a sum of external and internal factors, including cultural and perception transformations. Essential are new senses attributed by individuals to new or past landscapes (transformed or recreated); these connotations are related to collective representations, life stiles and spatial variations, interconnected and heightened by continuous contextual transformations (Popa 2011).

A photo of a certain space (understood as temporal frame captured from a landscape) can describe its basic elements that create atmosphere and character, but it

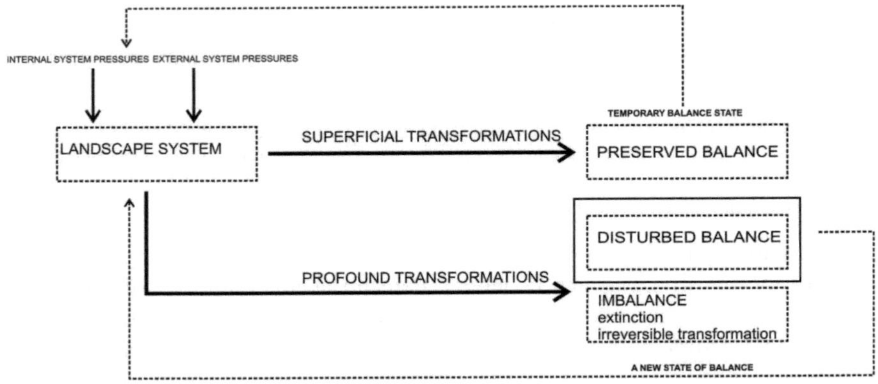

Fig. 6 Decoding landscape state trough photography interpretation (remote sensing)–assessment of different types of transformations undergone by constitutive elements

represents also a temporary state in landscape evolution. Accordingly, analysing time series photos, we can identify landscape change and transformation, but we can distinguish the same landscape due to preservation of its features. Opposite, sometimes we are confronted with the situation when landscape is not recognisable anymore due to alteration of its specific features or atmosphere, case in which the landscape tends to be ideal, just a reflective image of our consciousness about the place's lost characteristics. From this perspective, a landscape photo surprises and describes the place specificity, or creates a new perception, printed in our subconscious, based on pre-printed language that we use in order to decode the image of a place.

> The significant system of photography like classical painting, it is simultaneous a scene and a close look of the viewer, an object and a receiver (…) To the perspective, the representation system adds the frame (…) trough that frame, the world it is organised into a coherence which, in fact, is missing, in a paintings parade it is a sequence of decisive moments. Structure of representation—perspective and frame—is closely involved in reproducing ideology (Burgin 1982).

Retrieving the elements contained by a past landscape photo (lost landscape) over a decade can reveal a new image of a space and can lead to discover a new space or can reproduce a lost atmosphere, by reproducing an order of constitutive elements that does not exist anymore (Fig. 6).

5 Landscape Reconstruction, Recomposition and Revive Based on Photography: Conclusions

The article debates landscape photography as descriptor and temporal container of landscape features. By relating semiotic approach, photography as interpretation technique of landscape change and collective perception issues (codes), an

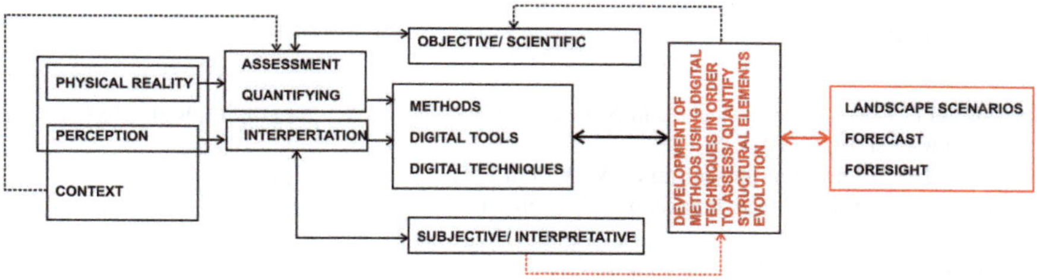

Fig. 7 Role of image interpretation (landscape decoding trough photography) in landscape dynamic/evolution assessment methodology

image (a photo) can describe space and atmosphere in many ways, starting from an iconic landscape, an invented one to a recognisable one.

The captured image of a totally transformed landscape (lost landscape) invites the receiver to use its own memories in order to decode the space, to find its characteristics, even if they are not anymore present. Also, photography can constitute a temporal and spatial journey, by transposing the viewer in an earlier stage of the evolving landscape.

Main landscape constitutive elements represented and with transformation possible to be decoded by photography are (Fig. 7):

- The limits/boundaries (expressed by fences, roads, rivers, green belts, paths) which can be modified (size, track), disappeared, similar.
- The district/patch (functional unit of landscape expressed by groups of buildings with similar features and structure, crops, agricultural plots) which can be similar or modified (size, limits, density, texture, disposal of elements, etc.).
- The landmark, offering the main characteristic of a place, giving it visibility and specificity: modified, disappeared and lost of features due to appearance of other elements.
- The path/corridor represented by communication flows (rivers, streets, green corridors and belts, etc.) contributing also to nodes transformations (constitutive elements) and matrix becoming and revealed by photography interpretation.

Space, place and landscape are concepts with different connotations for each individual. They represent a complex result of constitutive elements of a landscape (structural, formal, socio-cultural) and describe a certain stage of evolution (as temporary state). Constitutive elements are subject to continuous change and time transformation, and they have the capacity to mean or represent something by different intrinsic features: formal (spatial), historical significance, activities, habits, people, emotions, sensorial aspects. As a photo can capture all these characteristics, result is a description of a certain landscape state; also capturing some local, unique characteristics can induce and create specificity and atmosphere of a place (or reproduce it).

References

Bertrand G (1978) Le paysage entre la Nature et la Societe. Revue geographique des Pyrenees et du SudOuest 49

Burgin V (1982) Thinking photography. McMillan Press, New Jersey, pp 142–153

Cosgrove D (1994) Cultural landscapes. In: Unwin T (ed) A European geography. Longman, Edinborough Gate

Daniels S, Cosgrove D (1994) Introduction: iconography and landscape. In: Daniels S, Cosgrove D (eds) The iconography of the landscape: essays on the symbolic representations, design and use of past environments. Cambridge University Press, Cambridge

Forman RT, Gordon M (1986) Landscape ecology. Wiley, New York

Gruell, G. (1980). Fire's influence on wildlife habitat on the Bridger-Teton National Forest, Wyoming—volume I: Photographic record and analysis. Intermountain Region and Intermountain Forest and Range Experiment Station: Research paper int 235, 204 pp

Jackson JB (1984) Discovering the vernacular landscape. Yale University Press, New Haven

Kull C (2005) Historical landscape repeat photography as a tool for land use change research. Norsk Geografisk Tidsskrift—Norw J Geogr 5(59):253–268. Taylor Francis, London

Lynch K (1960) The Image of the City. Massachussettes Institute of Technology

Magill A, Twisse RH (1965) A guide for recording esthetic and biologic changes with photographs, Pacific southwest forest and range experiment station. US Forest service research note PSW 77, Berkley, Ref no. 907.2

Molyneaux B (1997) The cultural life of images. Routledge, London

Pocock DCD (1982) Valued landscape in the memory: the view form Prebend's Bridge. In: Transactions of the Institute of British Geographers, new series, vol 7, no. 3. Blackwell, Oxford, pp 354–364

Popa A (2011) Ipostaze ale peisajului agricol metropolitan (Aspects of agricultural landscape in metropolitan areas). In: Dabija AM (ed) *Spatiu urban- spatiu arhitectural- spatiu interior.* Universitara Ion Mincu, Bucharest, pp 83–93 (in Romanian)

Popa A (2012) Aspects of agricultural landscape as a cultural asset in metropolitan areas: case study for Bucharest city. Sci Future Lithuania 4(2):128–134

Sandu AL (2007) Under the sign of paradoxes. Foundation Arhitext Design, Bucharest

Part IV
Ecological Landscape: Cartography/Assessments/ Geography—Dynamics

Forest Landscape History Using Diachronic Cartography and GIS. Case Study: Subcarpathian Prahova Valley, Romania

Iuliana Armaş, Gabriela Osaci-Costache and Livioara Braşoveanu

Abstract The focus of this study is to assess the evolution of forest cover, as forest is an important indicator to describe the perceived naturalness of landscapes and has undergone radical changes in the last hundred years in Europe. The Subcarpathic sector of Prahova River was chosen as study area, because it has always been densely populated and its landscape has transformed by long-term man-made impact on the environment. Forest landscape spatial changes are highlighted in a temporal context based on large-scale historical maps and aerial images in a GIS environment. Different metrics are applied to quantify spatial changes of forest distribution and the pattern of changes over a 223 years period. The most important deforestations were recorded between 1790 and 1856 being correlated with the massive deforestation of Tara Romaneasca (the Southern Princedom of Romania), reported after the Treaty of Adrianople (1829) that liberalized trade and made wood an export product. An increase in forest cover took place between 1900 and 1980. Between 1980 and 1997, after the communist regime change (1989) and in the transition period, forest cover declined with the largest annual intensity, among causes are the restitution of forests to individual owners (followed by deforestation), legislative void, etc. Between 1997 and 2005, a slight tendency for reforestation was recorded. Comparison between 1790 and 2013 forest cover shows that almost 58 % of the landscape forest was destroyed, changing its use. Based on the index of naturalness, Prahova River Subcarpathic Valley has been included in the category of land insufficiently covered by forests for the entire period of analysis.

I. Armaş (✉) · G. Osaci-Costache · L. Braşoveanu
Faculty of Geography, Department of Geomorphology-Pedology-Geomatics, University of Bucharest, Bulevardul Nicolae Balcescu no. 1, Sector 1, 010041 Bucharest, Romania
e-mail: iuliaarmas@yahoo.com; iulia_armas@geo.unibuc.ro

G. Osaci-Costache
e-mail: gabrielaosaci68@yahoo.com; gabrielaosaci@libero.it

L. Braşoveanu
e-mail: brasoveanulivioara@yahoo.com

C. Crăciun and M. Bostenaru Dan (eds.), *Planning and Designing Sustainable and Resilient Landscapes*, Springer Geography, DOI: 10.1007/978-94-017-8536-5_6,

Keywords Forest cover changes • Diachronic analysis • Historical maps • Large-scale analysis • Aerial images • Binary change maps • GIS • Subcarpathians

1 Introduction

Rapid transformations of the landscape in the last centuries, particularly in the nineteenth and twentieth centuries, have implied the disappearance of 'remnants' of old landscapes, which can be recovered by diachronic analysis based on historical maps and aerial images of past periods. Historical cartography is a real 'treasure' to unearth and interpret (Rombai 2010, p. 69) and connect us to a time when the landscape was different from the one we see today. Analysis and comparison of historical maps are more important than that of historical documents, because it locates each landscape not only in time, but also in space. Currently, due to GIS software, data contained in old maps can be utilized in the digital environment (Azzari 2010) more effectively than by traditional cartography (Osaci-Costache 2002, 2004) giving a high precision to historical geographic studies (Rossi 2001; Osaci-Costache 2009a, b). Thus, retracing the ancient landscapes (both physical and cultural) helps understand the evolution of the environment under anthropic impact and establish an appropriate form of protection and territorial planning (Osaci-Costache 2009b, p. 107). There are numerous pieces of research demonstrating through case studies (Thomas 2001; Osaci-Costache 2004, 2011, 2012, etc.) or theoretically (e.g., Lago 2004; Rombai 2010) the importance of using historical maps. Since landscapes are an expression of an economic model, representing a mirror of civilization (Lorenzi 2007, pp. 145–146), in the same physical space, there are different types of landscape, which in time overlapped causing a real 'stratification' (De Vecchis 2004, p. 710). In this dynamic, among the most obvious phenomena that characterize territorial transformations are those relating to land use and especially forests (e.g., Osaci-Costache and Ene 2010; Pătru-Stupariu et al. 2011a, b, c; Pacheco Angulo et al. 2011; Popelkova and Mulkova 2011; Săvulescu and Mihai 2011; Skalos et al. 2012, etc.). In the Romanian Subcarpathian space, the extensive forests that existed at the end of the eighteenth century (and led to the formation of a civilization reliant on wood) are recording a dramatic decrease and only overlapping historical maps in GIS one can see the exact location, intensity and alternation of afforestation and deforestation.

2 Objective

In this study, we addressed the following questions:

1. How can we quantify spatial changes of forest distribution in a temporal context?
2. How can we identify the pattern of changes?

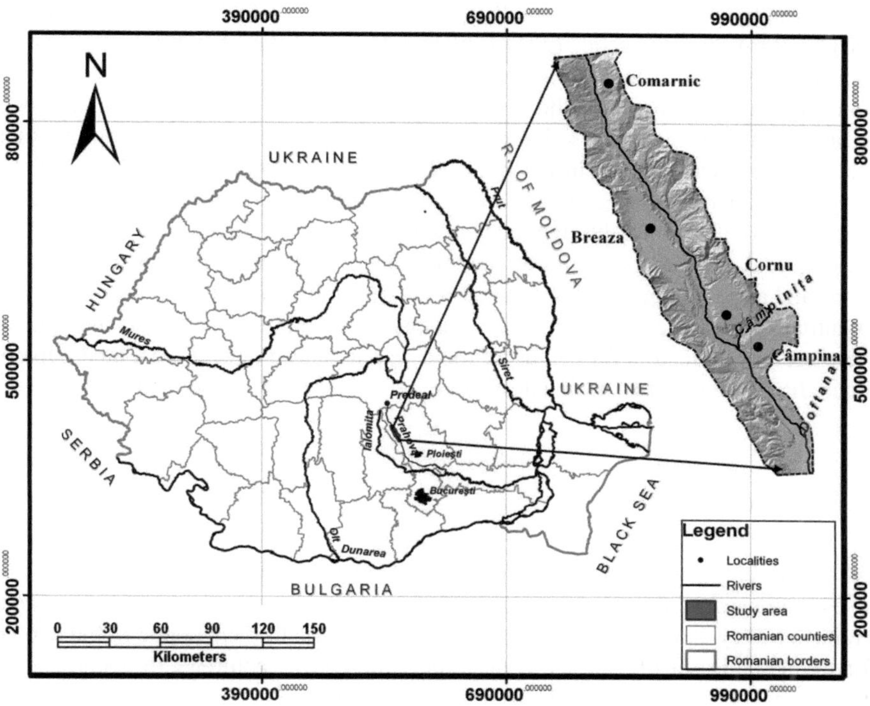

Fig. 1 Location of the Subcarpathian Prahova Valley

The case study area is the Subcarpathian Prahova Valley, situated between the town of Comarnic, upstream and the town of Câmpina, downstream (Fig. 1).

There are several reasons why we decided to study this area:

1. Firstly, the valley has been transformed by long-term human activities determining an increase in environmental changes. The increased landscape transformations due to human impact in the last centuries happened in the same time with a natural process of denudation caused by local geologic and geomorphologic factors (Armaş 2011a , b, 2013; Armaş et al. 2012, 2013);
2. Secondly, the area has been represented on good large-scale historical maps starting from the eighteenth century;
3. The Subcarpathic sector along Prahova River has always been densely populated with a density of about 180 inhabitants/km^2, being ranked the second most populated area of southern Romania after Bucharest capital city (Bugă et al. 1992). Proximity to the capital city and suitable natural environment—wide terraces and floodplains—provided optimal conditions for human settlement (Armaş et al. 2003; Vartolomei and Armaş 2010);
4. The Valley has been used for centuries as the only across Carpathian transport connection between Central and Southeastern Europe.

Recent landscape studies along the Prahova Valley were carried out by Pătru-Stupariu (2011), Pătru-Stupariu et al. (2011a, b) and Huzui (2012), but the authors either focused only on the mountain sector or considered all types of land use on shorter time intervals. The focus of this study is the evolution of forest cover, as forest is an important indicator to describe the perceived naturalness of landscapes and have undergone radical changes in the last 100 years in Europe (e.g., Tveit et al. 2006; Berg et al. 2008; Hultberg 2008). In our study, the category of forest cover is analyzed as a whole and not differentiated in scattered or dense forest.

The goal of our research was achieved by visual interpretation of historical maps and aerial images (orthophotomaps 2005, satellite images), change maps and quantitative measures in GIS aiming to identify evolutionary patterns. The metrics were used as indicators for capturing the dynamic nature of the forest landscape in a quantitative manner (Antrop and Van Eetvelde 2000; Dramstad et al. 2001).

3 Method and Data

For our research, we addressed the metrics as defined by Van Eetvelde (2007) to assess the spatiotemporal dynamics of forest landscapes along the Subcarpathian Prahova Valley. According to Van Eetvelde (2007), the first challenge in the choice for the suitable set of metrics to describe landscape structure is to acknowledge that every metric is unique. Thus, it has to be understood exactly what metrics have to be selected based on the purpose of the study (e.g., Botequilha Leitão and 2002; Fry 1996; Farina 1998, 2007). Secondly, sensitivity of metrics is essential. Would a change in metrics value equal a change in landscape? (Turner and Gardner 1991). Metric values are relevant only if they are comparable, across space and time with absolute reference values (Bogaert 2006). This last condition is more difficult to achieve, the more we focus on a category such as forest dynamics and return in the historical past. Historical and field research are necessary to define the driving forces that create the changes. However, because the objective of this case study is to identify spatial changes in forest cover and not to explain the driving forces of changes, we limited our research at the diachronic analysis of spatiotemporal changes based on large-scale historical maps in a period of 190 years, without considering other historical sources of information. For high accuracy of measurements on maps (and the calculations derived from them), we used only large-scale historical maps (Table 1a). The evolution of last decades was carried on through satellite images and orthophoto plans (Table 1b). The extent of the temporal scale is 223 years (1790–2013), and the grain is determined by six variable time intervals of 66, 44, 80, 17, 8 and 8 years. The grain of the temporal scale (the one given by above different variable time intervals) means that smooth oscillations occurring within a grain time interval cannot be shown in the analysis, being captured only the general trend in that time frame.

Nevertheless, the small amount of indices, the relative simplicity of the model used and the character of the input data (scale and resolution of analysis, the clear

Table 1 Overview of the data used in the analysis

a Historical maps

	Land surveys	Year of print	Scale	Average calculated residual error[c]	Transformation chosen	Digitization error
Specht's Map[a]	1790–1791	1790–1791	1:57,600	1,844.91 m (161.49 pixels)	Thin plate spline (TPS)	11.42 m (1 pixel)
Szathmary Map[a, b]	1855–1857	1864	1:57,600	48.60 m (19.72 pixels)	Thin plate spline (TPS)	2.46 m (1 pixel)
The Lambert's projection topographic plan[a]	1897–1902	1931–1955	1:20,000	8.10 m (4.76 pixels)	Helmert	1.70 m (1 pixel)
The Gauss–Krüger topographic map[a]	1980	1980	1:25,000	1.62 m (1 pixel)	Helmert	1.62 m (1 pixel)

b Aerial images used in GIS

Image record	Resolution(m)	Spectral bands combination used	Format	Data source
Landsat 7 ETM+ 1997	30	4-5-2	Raster	http://landsat.gsfc.nasa.gov
Orthophotomap 2005	5	Optical	Raster	SC Blom Romania SRL
Landsat 8 2013	30	4-5-2	Raster	http://landsat.gsfc.nasa.gov

[a] Georeferenced with Quantum GIS software under the reference system Dealul Piscului 1970/ Stereo 1970 (EPSG: 31700)
[b] Chart of southern Romania—copy at the scale of 1:57,600 of Marshal Fligely's map at the scale of 1/28,800
[c] Errors calculated by the Quantum GIS software for the Helmert transformation

borders of the study area and the physical borders) have not make it necessary to calculate additional statistical methods in selecting the metrics (e.g., correlation, factorial analysis). An overview of the data sources is given in Table 1a, b. For the georeference and superposition of 1790–1980 maps, we used QGIS open source software (version 2.0 Dufour, available at http://qgis.org/en/site/). Raster maps were obtained by scanning the images at a 400 dpi resolution and then georeferenced directly with QGIS. Helmert's spatial transformation was applied with the nearest neighbor resampling algorithm for the 1980 map and for the topographic plans under Lambert–Cholesky projection and with the thin plate spline transformation for Specht and Szatmary historical maps, which permit including some local deformations. For the latter ones, common ground control points were identified on the 1980 map. A large number of evenly dispersed points were selected (between 48 and 86 points/20 km^2, Armaş et al. 2012) in order to reduce the errors that may appear when working with historical maps, as pointed out by other authors (Dunn et al. 1990; Maffini et al. 1989). As ground control points (GCP), crossroads, train rails, bridges and churches were used (for Specht, GCP were bridges over rivulets, the interfluves line, important road junctions). The maps were brought to a common geo-cartographic referencing system (Dealul Piscului/Stereo 1970, EPSG: 31700), in order to allow successive superposition of forest evolution stages. Vector graphic layers (in ESRI Shapefile format) containing forested areas at different years were obtained in QGIS through digitization.

For identifying forested areas, on LANDSAT images, the supervised classification was applied, based on the combination of 5th, 4th and 2nd band in ENVI 4.7 software.

In a preliminary stage, we carried out overall forest change analysis by considering the statistics of the forested areas for each time layer. We estimated the absolute and percentagewise forest coverage for each layer together with the absolute and percentagewise changes between adjacent time layers.

To identify changes in adjacent time periods, we built the spatiotemporal database of different time layers and developed binary change maps for forest area evolution by using cross tabulation (e.g., Van Eetvelde and Käyhkö 2009; Van Eetvelde et al. 2012), respectively, transitions snapshoots between tx and tx + 1.

The binary change index is given by Van Eetvelde and Käyhkö (2009) as:

$$\mathrm{BCI} = \frac{\mathrm{NCH}\,\% \; - \mathrm{CH}\,\%}{\mathrm{NCH}\,\% \; + \mathrm{CH}\,\%} \tag{1}$$

where NCH % and CH % represent the forested area that remained unchanged and changed, respectively, during the time interval studied.

Because in this study we have not considered all types of land use, but we focused only on forest areas and their evolution, the binary change index was adapted accordantly to fulfill this goal. Unmodified and modified surfaces were not reported to the total area of the Subcarpathian Prahova Valley, but to the 'footprint' of the forest from the two successive historical maps analyzed. The 'footprint' (see Eq. 2) is the maximum extent of the forest (obtained by unifying the two vectorial layers) for each successive pair of maps (e.g., for 1790–1856, the forest footprint was calculated as the sum of 1790 forest surface and the new forest surface from 1856, without subtracting the areas transformed in another type of land use in that period of time).

$$\text{Forest 'Footprint'} = \mathrm{FA}_{\mathrm{Ti}} + \mathrm{AA}_{\mathrm{Ti-Tf}}, \tag{2}$$

Ti is the initial time (older map), Tf is the final time (newer map), $\mathrm{FA}_{\mathrm{Ti}}$ is the forest surface at $\mathrm{T}_{\mathrm{initial}}$ and $\mathrm{AA}_{\mathrm{Ti-Tf}}$ is the afforested surface between $\mathrm{T}_{\mathrm{initial}}$ and $\mathrm{T}_{\mathrm{final}}$ (between the two maps).

Thus, in the area so defined there are three types of situations that are important in the analysis of forest landscape dynamics: forest still remaining forest (unchanged use category), forest converted in a different type of land use and land converted to forest land use (the last two types are changed use category). We removed so the areas unoccupied by forests (unmodified in any way by the

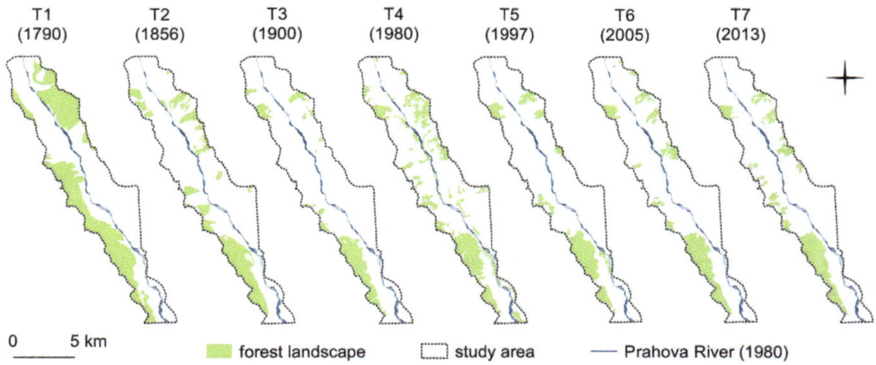

Fig. 2 Forested areas for every moment of analysis T (developed based on cartographic documents and aerial images)

appearance or disappearance of the forest), which are insignificant for the forest landscape dynamics.

Annual intensity of changes, which is the surface of deforestation or afforestation/year expressed in km^2/an, was calculated as a rate of forest cover change after the equation of Skalos et al. (2012):

$$I = \Delta P / \Delta N, \tag{3}$$

where I is the intensity of changes in forest cover, ΔP is the difference in the forest cover area between two adjacent time layers and ΔN is the number of years between two adjacent time layers.

4 Results

4.1 Changes in Forest Cover Between Time Layers

The forest cover maps of the analyzed time layers are shown in Fig. 2. Figure 3 gives an overview of the forest cover changes for each sequential time layer.

The largest forested area was mapped on the Austrian map from 1790, when forest represented over 35 % of the Subcarpathian Prahova valley, covering the entire slope areas (Figs. 2, 3). In 1834, the Prince Alexandru Ghica disposed that Campina custom is moved to Breaza de Sus setlement, where it will continue to function until 1852. In this period, Breaza de Sus settlement had developed becoming an important merchant crossing route between Muntenia and Transylvania. In 1856, perhaps due to economic growth and increase in trade in the area, the forest surface decreased by 18.76 % of the Subcarpathian Prahova Valley (about 47 % less than 1790 forest surface and with a decrease rate of 0.206 km^2/year).

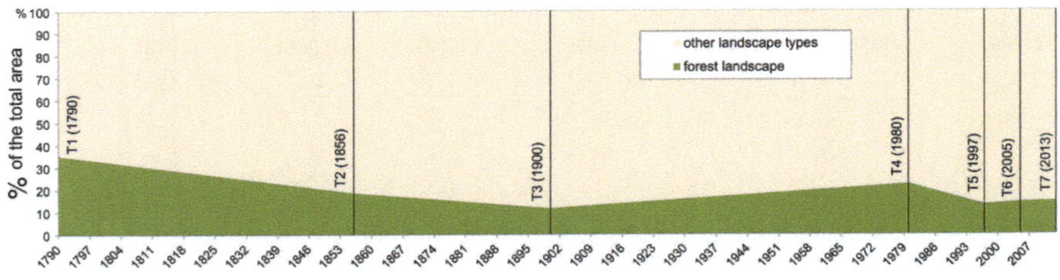

Fig. 3 Forest landscape dynamics in the Subcarpathian Prahova Valley based on cartographic data and aerial images (1790–2013)

Table 2 Forest cover changes between adjacent time layers and total change

Period	Increase/decrease in forest cover (km^2)	Increase/decrease in forest cover (%)
1790–1856	−13.641	−46.762
1856–1900	−5.756	−37.063
1900–1980	+8.964	+91.712
1980–1997	−7.683	−41.002
1997–2005	+0.901	+8.150
2005–2013	+0.304	+2.542
Total change 1790–2013	*−16.911*	*−57.971*

Oil exploitation (1860) and agricultural reform (1864) are other two possible reasons of further forest surface decrease by 11.81 % of the Subcarpathian Prahova Valley in 1900 (about 37 % less than in 1856, Table 2). The forested surface decrease in Prahova Valley correlates with the massive deforestation of Tara Romaneasca (Southern Princedom of Romania) reported after the Treaty of Adrianople (1829) ensuring freedom of trade (wood was an export product). 'Now logging becomes, as rightly said, devastating' (Giurescu 1976, p. 102).

An increase in forest cover took place between 1900 and 1980, with an annual average growth for the entire period of 0.11 km^2, a value identical to 1997–2005 (the largest growth rate for the 223-year period under review). In 1980, the forest represented about 23 % of the valley surface, a direct result of communist programs of reforestation—Law 204/1947 for the protection of forest patrimony, Law 3/1962 regarding the Sylviculture Code. Both laws provided for regeneration of logged areas, afforestations and reforestations, with strict rules regarding deforestation. In addition, Law 204/1947 provided afforestation for all agricultural lands (arable, meadow, pasture or orchards), located within the respective forests, owners of agricultural lands were given areas (an equal surface with the area they owned inside the forest) cleared on the edge of the forest through land exchange (cap. 3, art. 8), and Law 3/1962 recommended planting fast-growing wood species with high economic value (cap. 2, art. 15). Between 1980 and 1997, after the communist regime change (1989) and in the transition period, forest cover declined with the largest annual

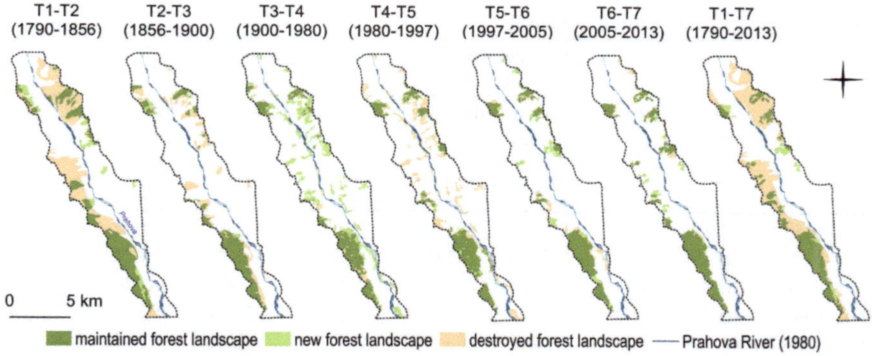

Fig. 4 Trajectories of changes (1790–2013)—developed based on cartographic documents

intensity. Among causes are the restitution of forests to individual owners (followed by deforestation), legislative void, etc. As a result, in 1997, the forest covered only 11 km^2, which is 13.35 % of the Subcarpathian Prahova Valley. During 1997–2013, the forest surface increased slightly up to 14.81 km^2, an increase by 0.112 km^2/year between 1997 and 2005 and 0.038 km^2/year between 2005 and 2013.

Comparing the situation from 1790 with 2013, we discovered that almost 58 % of the landscape forest was destroyed and its use transformed. The index of naturalness (forest percentage from total surface; Ionescu et al. 1989), calculated using cartographic data and aerial images classifies the study area in the category of land insufficiently covered by forests during the following years: 1856, 1900, 1997, 2005 and 2013, with values ranging between 11.81 % (1900) and 18.765 % (1856).

4.2 *Quantification of Changes*

In Table 2, there is the percentagewise aerial increase or decrease in the forest cover between adjacent time layers.

The transition and similarity matrixes and the binary change indices (Fig. 4; Tables 3, 4) show the direction of change (Baker 1989; Coppedge et al. 2007).

Analysis of Table 3 and the trajectories of changes (Fig. 4) show that from the end of eighteenth century to the middle of nineteenth century, 51.86 % of the forest surface was transformed in other land use categories with the highest value of forest cover change intensity (Table 5). Changes are concentrated especially in Breaza and Comarnic highland basins, marked by an intense phenomenon of 'swarming' and the formation of new settlements in deforested areas. In the same period, 94.24 % of other types of land use maintained their specific use, highlighting a traditional agricultural profile that expanded at the expense of the forest. The landscape forest that disappeared at the end of nineteenth century in Breaza highland basin records a slight increase as result of afforestation at the

Table 3 Transition and similarity matrix

Period	Forest 'footprint'		Unchanged use category: Forest still remaining forest		Changed use category: Forest converted in a different type of land use		Changed use category: Land converted to forest land use	
	km^2	%	km^2	%	km^2	%	km^2	%
1790–1856	32.258	100	12.442	38.57	16.729	51.86	3.087	9.57
1856–1900	16.754	100	8.55	51.033	6.98	41.662	1.224	7.305
1900–1980	20.179	100	8.333	41.295	1.441	7.141	10.405	51.564
1980–1997	20.272	100	9.515	46.937	9.223	45.496	1.534	7.567
1997–2005	14.069	100	8.942	63.558	2.113	15.019	3.014	21.423
2005–2013	13.823	100	10.393	75.185	1.563	11.308	1.867	13.507
1790–2013	*31.038*	*100*	*10.393*	*33.485*	*18.778*	*60.500*	*1.867*	*6.015*

Table 4 The binary change index of forest cover

Period	Unchanged (%)	Changed (%)	The binary change index
1790–1856	38.57	61.43	−0.228
1856–1900	51.033	48.967	+0.020
1900–1980	41.295	58.705	−0.174
1980–1997	46.937	53.063	−0.061
1997–2005	63.558	36.442	+0.271
2005–2013	10.392	3.430	+0.069
1790–2013	*33.485*	*66.515*	*−0.330*

Table 5 Forest cover change intensity

Period	Interval (years)	I = ΔP/ΔN (km^2/year)
1790–1856	66	−0.206
1856–1900	44	−0.130
1900–1980	80	+0.112
1980–1997	17	−0.452
1997–2005	8	+0.112
2005–2013	8	+0.038
Total change 1790–2013	*223*	*−0.076*

beginning of the twentieth century, specially on slope torrential basins and in floodplain (Table 2; Fig. 4).

The binary change index (Table 4) shows a higher change rate of wooded surfaces in the beginning of the oil exploitation in nineteenth century and during the construction of access routes toward Transylvania in twentieth century (the construction of the Câmpina-Sinaia railway: 1910–1912, the Bucureşti-Ploieşti-Câmpina-Sinaia-Braşov national road built from 1964 to 1965 continuing to be modernized until the construction of the present route located on the left bank of the river and parallel to the Prahova Valley).

In the post-communist era, especially until 2000, the areas reforested in the 1980s are again cleared. A slight reforestation of these areas, largely susceptible to accelerated erosion, is noted after 2005. The forest continuity analysis during the entire study period shows that forests are kept constant on the Subcarpathian slopes of the right side of the river, with an increased stability especially south of Campina Town as a result of the accentuated energy of cuesta relief and the presence of molasses deposits formed in the Miocene–Pliocene–Lower Pleistocene interval (Fig. 4). Poorly cemented sands, sandstones alternating with marly clays and clays maintain current active processes on steep slopes, even under the forest cover. In this sector, over 33 % of the 1790 forest area is maintained as forest up to present times.

5 Conclusions

The analysis of the spatial distribution of forests obtained from historical maps and aerial images provided useful information about the changes of forested areas along the Subcarpathian Prahova Valley. The main changes in forest distribution were identified in the 1790–1856 period, after the signing of Adrianople Treaty (1829) which liberalized trade, making wood an export product. Another massive deforestation period was marked by the change of regime in 1989 until 2000, when the retrocession of forests and the legal void resulted in significant deforestation. Throughout the period 1790–2013, although there were short periods of reforestation, almost 58 % of the Subcarpathian Prahova Valley forest landscape was destroyed. Forest was maintained constant on the steep and fragmented slopes with high instability and affected by landslide processes. The area has been highly favored for human settlement since ancient times and this fact is reflected in the high degree of transformation of the natural landscape. The index of naturalness classifies this hilly area in the category of land insufficiently covered by forests.

Acknowledgments This chapter is partially based upon work developed within the research project 2916/31GR, financed by the National Research Council (CNCS) of Romania and having Prof. Iuliana Armaş as Principal Investigator.

References

Antrop M, Van Eetvelde V (2000) Holistic aspects of suburban landscapes: visual image interpretation and landscape metrics. Landscape Urban Planning 50(1–3):43–58

Armaş I, Damian R, Şandric I, Osaci-Costache G (2003) *Vulnerabilitatea versanţilor la alunecări de teren în sectorul subcarpatic al văii Prahova*. Fundaţiei România de Mâine, Bucureşti

Armaş I (2011a) An analytic multicriteria hierarchical approach to assess landslide vulnerability. Case study: Cornu Village/Romania. Z Geomorphol 55(2):209–229

Armaş I (2011b) Weights of evidence method for landslide susceptibility mapping. Prahova Subcarpathians, Romania. Nat Hazards 60(3):937–950

Armaş I, Gogoase Nistoran D, Osaci-Costache G, Brasoveanu L (2012) Morpho-dynamic evolution patterns of Subcarpathian Prahova Valley (Romania). Catena 100:83–99

Armaş I (2013) Diagnosis of landslide risk for individual buildings. Insights from Prahova Subcarpathians, Romania. Environ Geol. doi:10.1007/s12665-013-2854-5 (in press)

Armaş I, Vartolomei F, Stroia F, Brasoveanu L (2013) Landslide susceptibility deterministic approach using geographical information systems: application to Breaza city, Romania. Nat Hazards 1–23 (in press)

Azzari M (2010) Prospettive e problematiche d'impiego della cartografia del passato in formato digitale. Bollettino dell'Associazione Italiana di Cartografia 138:217–224

Baker WL (1989) A review of models of landscape change. Landscape Ecol 2:111–133

Berg AL, Ostlund MJ, Olofsson J (2008) A century logging and forestry in a reindeer herding area in northern Sweden. For Ecol Manage 256:1009–1020

Bogaert J (2006) On the definition and use of landscape metrics to disentangle landscape complexity. In: Brancucci G (ed) Pianificare l'incertezza. L'ecologia del paesaggio nella gestione dei sistemi territoriali. IALE, Genova, pp 11–24

Botequilha Leitao A, Ahern J (2002) Applying landscape ecological concepts and metrics in sustainable landscape planning. Landscape Urban Planning 59:65–93

Bugă D, Baranovski N, Muică N (1992) Subcarpaţii Prahovei. In: Geografia României. Regiunile pericarpatice: Dealurile şi Câmpia Banatului şi Crişanei, Podişul Mehedinţi, Subcarpaţii, Piemontul Getic, Podişul Moldovei. Editura Academiei Române, Bucureşti, pp 276–291

Coppedge BR, Engle DM, Fuhlendorf SD (2007) Markov models of land cover dynamics in a southern Great Plains grassland region. Landscape Ecol 22:1383–1393

De Vecchis G (2004) Denominazioni comuni e nomi propri di località abitati. Atlante dei tipi geografici. Istituto Geografico Militare, Firenze, pp 710–714

Dramstad WE, Fry G, Fjellstad WJ, Skar B, Helliksen W, Sollund MLB, Tveit MS, Geelmuyden AK, Framstad E (2001) Integrating landscape-based values—norwegian monitoring of agricultural landscapes. Landscape Urban Planning 57:257–268

Dunn R, Harrison AR, White JC (1990) Positional accuracy and measurement error in digital databases of landuse: an empirical study. Int J Geogr Inf Syst 4:385–398

Farina A (1998) Principles and methods in landscape ecology. Chapman & Hall, London

Farina A (2007) Principles and methods in landscape ecology, towards a science of landscape. Springer, New York, p 412

Fry G (1996) A landscape perspective of biodiversity; indices, models and planning. In Simpson IA, Dennis P (eds) The spatial dynamics of biodiversity. Proceedings of the 5th annual IALE (UK) conference, Stirling, p 3–16

Giurescu CC (1976) Istoria pădurii româneşti din cele mai vechi timpuri până astăzi. Editura Ceres, Bucureşti

Hultberg T (2008) Forest continuity and human impact-vegetation history of Torup forest, south–western Scania. Examensarbete 107, Swedish University of Agricultural Sciences, Institutionen fo¨ r sydsvensk skogsvetenskap, Alnarp

Huzui A (2012) Analiza cantitativă şi calitativă a peisajului urban. Studiu de caz: oraşul Sinaia, teza de doctorat (rez.), Şcoala doctorală "Simion Mehedinţi", Facultatea de Geografie, Universitatea din Bucureşti, Bucureşti

Ionescu A, Săhleanu V, Bîndiu C (1989) Protecţia mediului înconjurător şi educaţia ecologică. Editura Ceres, Bucureşti

Lago L (2004) Il contributo della cartografia storica. Atlante dei tipi geografici. Istituto Geografico Militare, Firenze, pp 21–27

Lorenzi R (2007) La stratigrafia paesaggistica: proposta di metodo per la mappatura dei paesaggi storici. In Persi P (ed) Recondita armonia. Il paesaggio tra progetto e governo del territorio. Istituto Interfacoltà di Geografia, Università degli Studi di Urbino "Carlo Bo", Urbino, p 145–152

Maffini G, Arno M, Bitterlich W (1989) Observations and comments on the generation and treatment of error in digital GIS data. In: Goodchild M, Gopal S (eds) The accuracy of spatial data bases. Taylor and Francis, London, pp 55–69

Osaci-Costache G (2002) Cartografierea dinamicii peisajului geografic din zona subcarpatică dintre Dâmboviţa şi Olt reflectată în documentele cartografice. Teză de doctorat, Facultatea de Geografie, Universitatea din Bucureşti, Bucureşti (mss.)

Osaci-Costache G (2004) Muşcelele dintre Dâmboviţa şi Olt în documente cartografice. Reconstituirea şi dinamica peisajului geografic în secolele XVIII–XX. Editura Universitară, Bucureşti

Osaci-Costache G (2009a) Cartografia storica in ambiente GIS Open Source (Quantum GIS e gvSIG) per l'analisi delle dinamiche dell'utilizzazione dei terreni: il caso della città di Curtea de Argeş nel Novecento. Analele Universităţii Ştefan cel Mare Secţiunea Geografie 18:233–242

Osaci-Costache G (2009b) L'applicazione della cartografia retrospettiva agli studi territoriali: vecchi paesaggi rurali nello spazio subcarpatico ubicato tra i fiumi Topolog e Doamnei (Romania) applying the retrospective cartography in the territorial studies: old rural landscapes in the subcarpathian space between the topolog and doamnei rivers (Romania). In Persi P (ed) Territori contesi. Campi del sapere, identità locali, istituzioni, progettualità paesaggistica, IV Convegno Internazionale Beni Culturali, Pollenza, Italia. Istituto Interfacoltà di Geografia, Università degli studi di Urbino "Carlo Bo", pp 107–112

Osaci-Costache G (2011) Cartografia e paesaggi del passato in ambiente GIS Libero e Open Source: il caso della zona di contatto tra i Monti Cozia e la depressione subcarpatica Jiblea-Berislăveşti (Romania). In: Gregori L (ed) Il paesaggio tra reale e virtuale. Nuova Prhomos, Città di Castello, pp 337–348

Osaci-Costache G (2012) Cartografia storica e GIS Open Source per la valutazione dell'evoluzione del paesaggio dovuta all'impatto antropico. Il caso del bacino subcarpatico Cicăneşti (Romania)/Historical cartography and GIS open source for the assessment of landscape evolution due to anthropogenic impact. Case study: the Cicăneşti Subcarpathian basin (Romania). Bollettino dell'Associazione Italiana di Cartografia 143:61–76

Osaci-Costache G, Ene M (2010) The analysis of forest dynamics within the Carpathians—the Subcarpathians contact area by using the historical cartography approach and open source GIS software. Case study: the Limpedea catchment (Romania). Geogr Phorum. Geogr Stud Environ Prot Res 9(9):115–124

Pacheco Angulo C, Aguado Suárez I, Mollicone D (2011) Dinámica de la deforestación en Venezuela: análisis de los cambios a partir de mapas históricos. Interciencia 36(8):578–586

Pătru-Stupariu I (ed) (2011) Elaborating and implementing an algorithm for landscape evaluation and prognosis. Application for the mountainous and the subcarpathian sector of Prahova Valley. Univ. din Bucureşti, Bucureşti

Pătru-Stupariu I, Stupariu MS, Cuculici R, Huzui A (2011a) Contribution of global indicators to landscape change modelling. Case study: Prahova Valley (Romanian Carpathians and Subcarpathians). Int J Phys Sci 6(3):534–539

Pătru-Stupariu I, Stupariu MS, Huzui A (2011b) Application of landscape metrics to assess the interaction between nature parks and their urban surroundings. Case study: the Bucegi Nature Park-Sinaia city fringe, I.F.L.A. World Congress, Zurich, June 2011

Pătru-Stupariu I, Stupariu MS, Cuculici R, Huzui A (2011c) Understanding landscape change using historical maps. Case study Sinaia, Romania. J Maps 7(1):206–220. doi:10.4113/jom.2011.1151

Popelkova P, Mulkova M (2011) Landscape changes mapping: central part of Ostrava-Karvina mining district, Czech Republic. J Maps 2011:363–375. doi:10.4113/jom.2011.1165

Rombai L (2010) Le problematiche relative all'uso della cartografia storica. Bollettino dell'Associazione Italiana di Cartografia 138:69–89

Rossi A (2001) Cartografia storica e GIS: proposte per la pianificazione del territorio. Bollettino dell'Associazione Italiana di Cartografia 477–492

Săvulescu I, Mihai B (2011) Mapping forest landscape change in Iezer Mountains, Romanian Carpathians. A GIS approach based on cartographic heritage, forestry data and remote sensing imagery. J Maps 2011:429–446. doi:10.4113/jom.2011.1170

Skalos J, Engstova B, Trpakova I, Santruckova M, Podrazsky M (2012) Long-term changes in forest cover 1780–2007 in central Bohemia, Czech Republic. Eur J Forest Res 131:871–884. doi:10.1007/s10342-011-0560-y

Thomas MF (2001) Landscape sensitivity in time and space—an introduction. Catena 42:83–98
Turner MG, Gardner RH (eds) (1991) Quantitative methods in landscape ecology. Springer, New York, pp 77–103
Tveit M, Ode A, Fry G (2006) Key concepts in a framework for analysing visual landscape character. Landscape Res 31(3):229–255
Van Eetvelde V (2007) Van geografische strekenkaart tot landschapsdatabank. Gebruik van GIS, informatietheorie en landschapsmetrieken voor het karakteriseren van landschappen, toegepast op België. Universiteit Gent, Vakgroep Geografie, Gent. Promotor: Prof. Dr. Marc Antrop
Van Eetvelde V, Käyhkö N (2009) The applicability of quantitative techniques for assessing spatio-temporal patterns of landscape changes. In Breuste J, Kozová M, Finka M (eds) Proceedings of the European IALE conference 2009, pp 379–382
Van Eetvelde V, Antrop M, Van de Velde L (2012) Assessing the actuality of historical maps in rural landscapes, case studies of settlements in Belgium. In: 15th international conference of historical geographers, Prague
Vartolomei F, Armaş I (2010) The intensification of the anthropic pressure through the expansion of the constructed area in the subcarpathian sector of the Prahova Valley/Romania (1800–2008). Forum Geografic 9(9):125–132

GIS-Based Methodology for the Analysis of Regional Landscapes and Their Changes Based on Land Cover and Use: A Planning Perspective Aimed at Conserving the Natural Heritage

Alexandru-Ionuţ Petrişor

Abstract Considered by the ecologists a complex of ecosystem and by the geographers a territorial system, the landscape represents the result of long-term interaction between man and nature, translated into a certain specific structural and functional organization and into the way it is perceived by human communities. The interaction between man and landscape is reflected in land cover and use changes, which represent an instrument to assess trends of interest for planning purposes. Moreover, climate changes could determine additional land cover and use changes, which at their turn modify the landscape. This chapter proposes a methodology using the geographical information systems (GIS) and CORINE land cover and use data to assess the current environmental issues and trends and determine the priorities of planning and strategies of development, starting from the example of Covasna County in Romania.

Keywords Biodiversity • Natural protected sites • Real estate boom • Global change • Agriculture • Urbanization

A.-I. Petrişor (✉)
Assistant Professor / Lecturer, "Ion Mincu" University of Architecture and Urbanism, Str. Academiei nr. 18-20, Sector 1, 010014 Bucharest, Romania
e-mail: alexandru_petrisor@yahoo.com

A.-I. Petrişor
Scientific Director for Urban Planning and Territorial Development and Senior Researcher, National Institute for Research and Development in Constructions, Urbanism and Sustainable Spatial Development URBAN-INCERC, sos. Pantelimon nr. 266, Sector 2, 021652 Bucharest, Romania

C. Crăciun and M. Bostenaru Dan (eds.), *Planning and Designing Sustainable and Resilient Landscapes*, Springer Geography, DOI: 10.1007/978-94-017-8536-5_7,

1 The Concept of "Landscape": A Multi-Disciplinary Approach

Sustainable development, defined by the Brundtland report based on meeting at the same rate current and future needs (Brundtland 1987), presumes that natural resources are used within the limits of the carrying capacity, biodiversity is preserved, deteriorated systems are restored, and environmental protection is embedded in sectoral development strategies, in order to internalize environmental costs and assess the environmental impact. All these issues rely on a holistic understanding of the environment, perceived as a whole consisting of the socioeconomic and natural capitals. Many authors acknowledged three traditional pillars of sustainability—economic, social, and environmental (Basiago 1999; Littig and Grießler 2005; Gibson 2006; Murphy 2012). However, other authors add a fourth cultural one, acknowledging its potential for economic growth (Hawkes 2001). Its recognition was resulted from an intense lobby by the United Cities and Local Governments, who recognized it formally in 2010 (United Cities and Local Governments 2010). Therefore, cultural diversity is seen as a component of biodiversity.

The documents of the European Conference of Ministers Responsible for Regional/Spatial Planning (CEMAT) show that sustainable spatial development ensures the coherence of socioeconomic targets with respect to the territory and its ecological and cultural functions, aiming to enhance the quality of life for current and future generations by the creation of sustainable communities able to manage and use the resources efficiently, exploiting the ecological and social potential of the economy for innovation and safeguarding the welfare, environmental protection, and social cohesion (Collignon 2009).

In geography, Carl Troll (1968) defined "landscape ecology" as the study of morphology, classification, time change, and functional relationships within different geographical units; few years later, Isaak S. Zonneveld (1972) defined the landscape as "*holistic entity formed by different element influencing each other*" (Wu and Hobbs 2007). In ecology, "landscape" is used in the literature to define both the landscape and complex of ecosystems. Prof. Dr. Angheluţă Vădineanu considers that systemic ecology provides the theoretical background required for conceptualizing and understanding the structure and functions of coupled complexes of ecological systems, either social or natural (Vădineanu 2004). Thus, the object of ecology changed, since its original definition (Haeckel 1866) from organisms to biotic communities (biocoenoses)—Andrewartha and Birch (1954), then ecosystems (Tansley 1935; Pickett and Grove 2009), and finally complexes of ecosystems (Vădineanu 2004). All levels are generalized by the concept of ecological system (Vădineanu 1998). In spatial planning, landscapes are defined as units formed by the interaction between man and nature (Parliament of Romania, 2002) across time (IUCN 1994) and its perception by the population (Parliament of Romania 2002; Philips 2002).

A distinct category of reserves is represented by the *protected terrestrial or marine landscapes* (category V of the International Union for Conservation of Nature) or *natural parks* (Parliament of Romania 2007), defined as *area of land, with coast and sea as appropriate, where the interaction of people and nature over time has produced an area of distinct character with significant aesthetic, ecological and/or cultural value, and often with high biological diversity; safeguarding the integrity of this traditional interaction is vital to the protection, maintenance and evolution of such an area* (IUCN 1994). Commenting this definition, the Slovenian Ministry of the Environment and Spatial Planning shows that "*nature and culture come together in the landscape*" (Jančič 2007). Within the same conceptual framework, the *cultural landscape* is defined by the de Council of Europe Framework Convention on the Value of Cultural Heritage for Society (Faro 2005) as *group of resources inherited from the past which people identify, independently of ownership, as a reflection and expression of their constantly evolving values, beliefs, knowledge and traditions; it includes all aspects of the environment* ***resulting from the interaction between people and places through time*** (COE 2005). The document underlines the importance of preserving this heritage in order to achieve a sustainable development.

All the aforementioned conceptual models can be reunited in a definition used in the landscape classes at "Ion Mincu" University of Architecture and Urbanism in Bucharest, reflecting the vision of the Landscape Department: *The landscape represents a part of a territory as seen, which in its natural state and/or by human intervention constitutes an aesthetic assembly representing a subjective reality, with a spatial dimension and a temporal one, both under continuous change. It is a type of synthesis resulting from the interaction between nature and culture, an "archive document" and a "fingerprint" of the interaction; it is the reflection, perception and history of a society, but also the aesthetic relationship between man and the environment, revealed by a cultural construction* (Crăciun 2010).

All these elements seem to indicate that the landscape is the reference study unit in geography, ecology, and spatial planning. Moreover, other disciplines—including sciences and arts—use this concept as well. Even though its interpretation is different, there are common elements, as the territorial reality is the same.

Jensen (2000) defines land cover as "biophysical materials found on the land" and land use as "how the land is being used by human beings." However, Petrişor et al. (2010) make a difference between natural and artificial systems, considering that land use provides a detailed classification of natural systems and indicates the utilization of the artificial ones by human communities. The development of human communities results in environmental changes, reflected ultimately by changes in land cover and use. Provided that space is limited in general, and the space usable by human is even more limited in particular, there is a clear spatial conflict between man-dominated systems and the natural ones. Adverse impacts of socioeconomic development, amplified when the process is not controlled, include urban sprawl and unwise land use, with consequences that affect not only the natural systems, but also those who generated them, i.e., human communities.

In addition, land cover and use are described by CORINE land cover and use classification since land cover is described by its first level (CLC1) and land use by the next ones (CLC2 and 3), depending on the resolution of analysis

(Petrişor et al. 2010). In addition, Petrişor (2008) connects the levels of CORINE classification to the Nomenclature of Territorial Units for Statistics and to the spatial levels of biological diversity, indicating which CORINE levels are appropriate to analyze each type of administrative unit from an environmental standpoint, as part of the planning process. This is particularly important since the changes in land cover and use have as main underlying cause socioeconomic and political transformations (Petrişor et al. 2010).

All the theoretical elements described above can substantiate a methodology for analyzing the environmental processes, characterizing the current situation and establishing planning goals for the future. The methodology employs extensively the usage of geographical information systems (GIS), understood as "decision support systems involving the integration of spatially referenced data in a problem solving environment" (Cowen 1988). The option for GIS is justified by their analytical power, but also by the ability to derive spatial relationships (and make subsequent computations) combining information from different datasets to describe the biophysical structure of the territory and look at the trends manifested by its modifications under the human pressure. The approach can substantiate better urban and spatial plans, but has also a high potential for research, due mainly to its quantitative techniques. The potential is amplified when the method is used in conjunction with other techniques.

2 Data and Processing

Data used with GIS in the case study presented in the next section come mainly from international sources; Table 1 summarizes all datasets, listing the provider, format, and transformations. The European Environment Agency (EEA) provides regularly data on land cover and use and biogeographical regions. Nevertheless, generation of data covering an entire continent is possible at large time intervals. Available data start from 1990, and the next reference years are 2000 (data available in 2004) and 2006 (available in 2010). Clime data are freely available from the University of Berkeley in a DIVA-GIS format (Hijmans et al. 2001). Current data are an output of the WorldClim project (Hijmans et al. 2005), and 2100 predictions are based on double CO_2 concentrations and the CCM3 model (Govindasamy et al. 2005). The Romanian natural protected areas of national importance data are offered by the Romanian Ministry of the Environment and Forests free of any charge. The World Wildlife Fund provides data on the terrestrial ecological regions free of charge.

The data required spatial processing (Table 1), such as change of format and/or projection, clipping for the national territory. For data in an original raster format (elevation, clime), a special methodology was built to transform it into a vector format: Each cell was reduced to its center, and centers were interpolated based on some value (elevation, temperature, etc.) using radial basis functions, to derive smooth surfaces. The areas were reclassified to acquire a certain resolution, and filled contours were exported in a vector format. The applicability of the method depends on the size of the administrative unit and the original raster resolution.

Table 1 Data: sources and processing

Dataset	Provider	Format	Remarks	Transformations
Climate—actual	University of Berkeley[a]	DIVA-GIS	Produced by the project WorldClim; 2.5 min × 2.5 min	Imported in ArcView GIS 3.X, project into Stereo 1970, subsample for Romania
Climate—predicted	University of Berkeley[b]	DIVA-GIS	Predictions for 2100 based on 2 × CO_2 concentration and CCM3 model; 2.5 min × 2.5 min	Imported in ArcView GIS 3.X, project into Stereo 1970, subsample for Romania
Biogeographical regions	European environment agency—EEA[c]	ArcView GIS 3.X	Newest version	Project into Stereo 1970, subsample for Romania
Ecoregions	World wildlife fund[d]	ArcView GIS 3.X	Newest version	Project into Stereo 1970, subsample for Romania
Elevation	Consulting group on international agricultural research[e]	DEM	Nearly 90 m × 90 m	Import into Arc GIS, then export to ArcView GIS 3 X, project into Stereo 1970, subsample for Romania
Land cover and use	EEA[f]	ArcView GIS 3.X	2006 data	Project into Stereo 1970, subsample for Romania
Land cover and use changes	EEA[f]	ArcView GIS 3.X	1990–2000 and 2000–2006 data	Project into Stereo 1970, subsample for Romania
Protected natural sites	Romanian Ministry of the Environment[g]	ArcView GIS 3.X	Only sites of national importance included	None

[a] http://biogeo.berkeley.edu/worldclim/diva/diva_worldclim_2-5m.zip
[b] http://biogeo.berkeley.edu/worldclim/diva/diva_wc_ccm3_2-5m.zip
[c] http://www.eea.europa.eu/data-and-maps/figures/biogeographical-regions-in-europe
[d] http://www.worldwildlife.org/science/ecoregions/item1267.html
[e] http://srtm.csi.cgiar.org/SELECTION/inputCoord.asp
[f] http://dataservice.eea.europa.eu/dataservice/metadetails.asp?id=650
[g] http://www.mmediu.ro/departament_ape/biodiversitate

Fig. 1 Position of Covasna County (*red*) in Romania and Europe

2.1 Case Study: Covasna County, Romania

Covasna County, covering approximately 3,705 km^2, is a Romanian county situated in Transylvania, in the southeast of the center region of development, adjacent to the counties Vrancea, Bacău, Harghita, Braşov, and Buzău (Fig. 1). Elevations range between 468 and 1,177 m (Fig. 2). Peaks are situated in east and north and lower elevations in the center. Biogeographically (Fig. 3), it is covered partially by the Alpine region (south, east and north) and the continental region (west). The same pattern characterizes its ecological regions (Fig. 4), with Carpathian mixed forests in the mountain areas (south, east and north) and Pannonian mixed forests at lower elevations, in the west. The configuration of relief is reflected by climate, with higher temperatures (Fig. 5) and lower precipitations (Fig. 6) at lower elevations.

Climate predictions exhibit an overall similar, but different in details pattern. Temperatures are expected to increase in the middle part, from east to west, affecting mostly the mountain and hill regions (Fig. 7), and precipitations tend to decrease from south to north (Fig. 8).

Land cover is also a reflection of the relief. Mountain areas are covered with forests; human settlements are developed at lower elevations, in the depressions, and are surrounded by agricultural areas. Agricultural areas and forests make up most of the total surface. There are few waters (streaming or stagnant) and wetlands in the center (Fig. 9). Land use is presented in Fig. 10 and detailed in Table 2. The most important areas are forests (41 %) and agricultural areas (54 %). Human settlements represent only 4 %. Among the agricultural areas, the most important are non-irrigated arable land (21 %) and pastures (14 %), and among natural areas, broad-leaved forests (28 %), coniferous forests (9 %), and mixed forests (11 %). The predominant categories are also the most affected by land cover and use changes (Fig. 11).

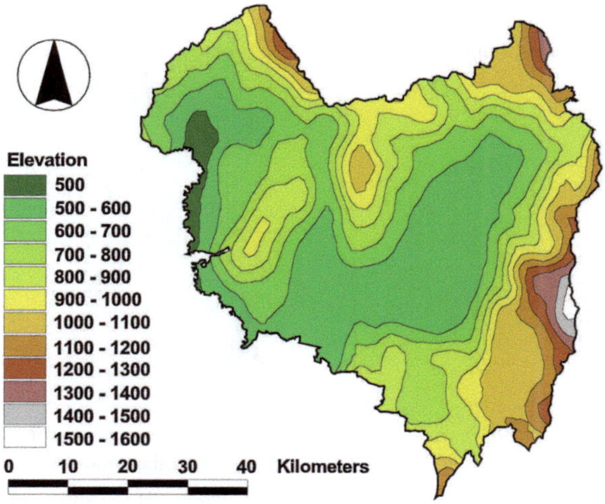

Fig. 2 Elevation in Covasna County

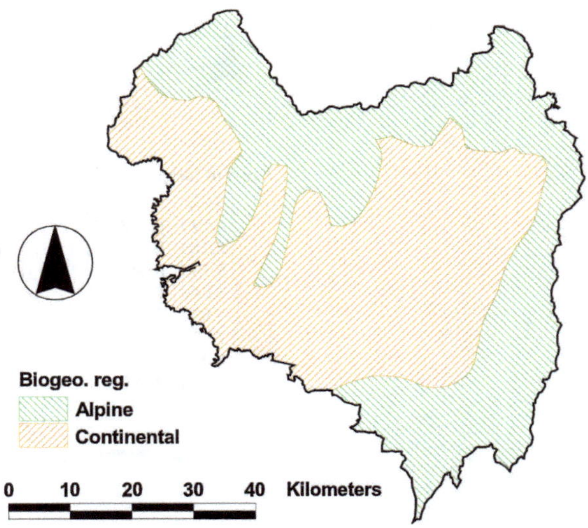

Fig. 3 Biogeographical regions in Covasna County

Particularly during 1990–2000, massif portions of the forests were cut off after returning to the families of former owners. In the new economic context, this led to a true drama (Roman 2009). Nevertheless, forests were also replanted, in a different area. In fact, the regeneration of forests could be due to two phenomena, which cannot be distinguished without knowing the field reality: Afforestation is

Fig. 4 Ecological regions in Covasna County

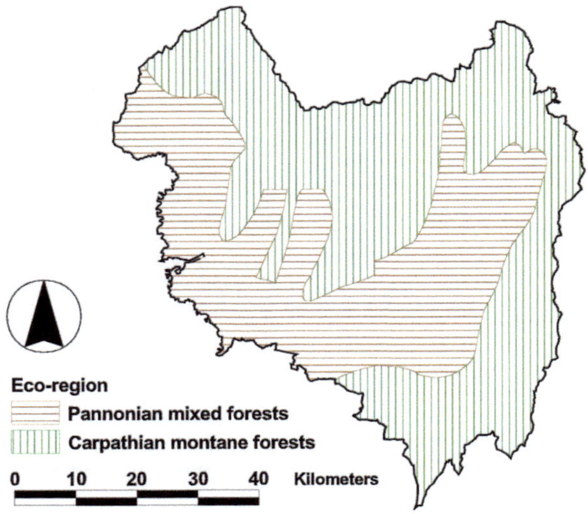

Fig. 5 Current temperatures in Covasna County

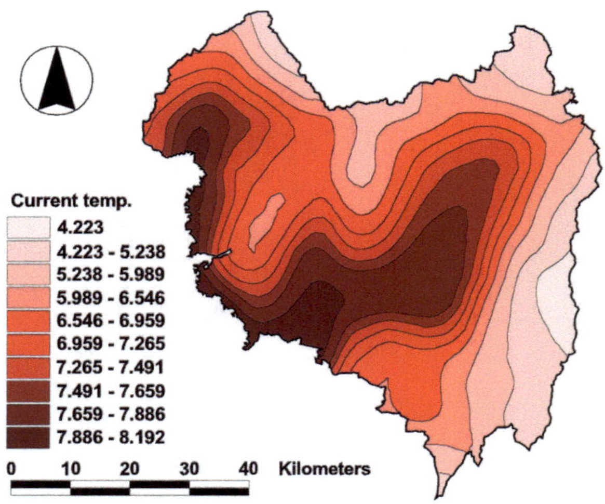

a conversion from other land uses into forest, or increase of the canopy coverage to above the 10 % threshold, achieved through plantations or natural regeneration, while reforestation is the re-establishment of forest formations after a temporary condition with less than 10 % canopy cover due to human-induced or natural perturbations (Dutcă and Abrudan 2010). If most forest cuts occurred in the northeast and east, they grew in north–northeast and east–northeast. Two other important processes were the development and abandon of agriculture. While the development of agriculture can easily be located in the northeast, its abandon is scattered throughout the entire territory.

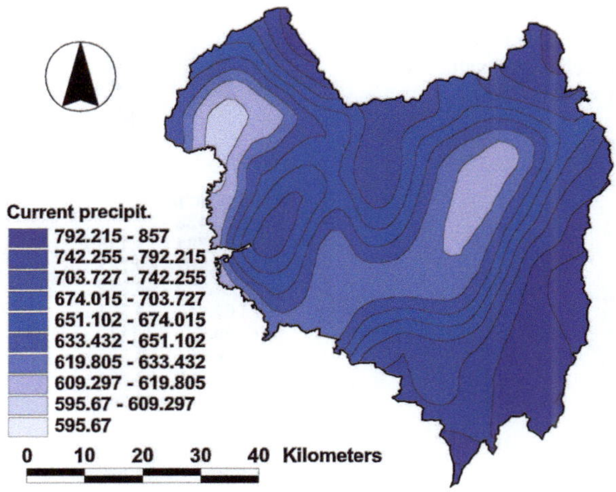

Fig. 6 Current precipitations in Covasna County

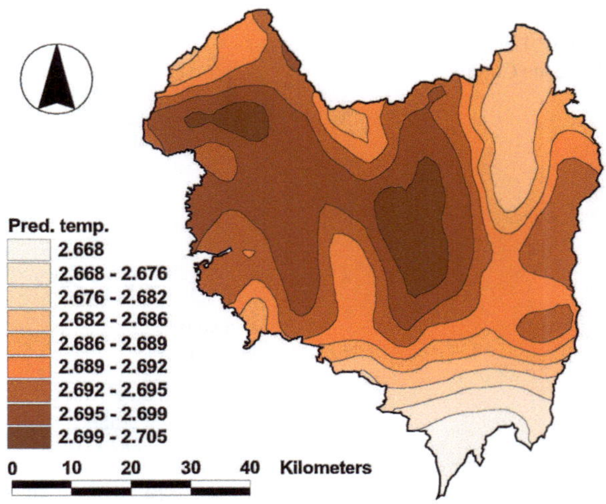

Fig. 7 Differences between current and predicted temperature values in Covasna County

Table 3 presents the figures related to land cover and use. During 1990–2000, deforestation affected nearly 10,000 ha, making up 57 % of total land cover and use changes. In compensation, forestation (natural or anthropic) restored 27 %. Therefore, the difference (30 % of all land changes) indicates the magnitude of forest loss. Similarly, given that the abandon of agriculture makes up 9 % of changes, its development represents only 7 %, the difference of 2 % accounting for the loss of agricultural potential.

During 2000–2006, deforestations continue, representing 40 % of land cover and use changes, but the effects of the real estate boom become visible, and urbanization accounts for 800 ha, meaning 48 % of all changes.

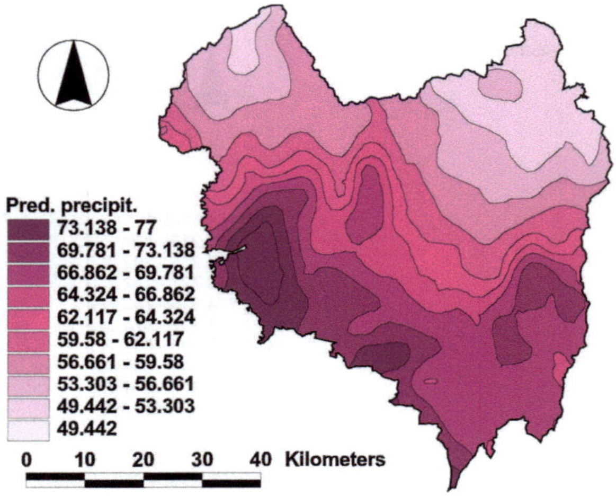

Fig. 8 Differences between current and predicted precipitation values in Covasna County

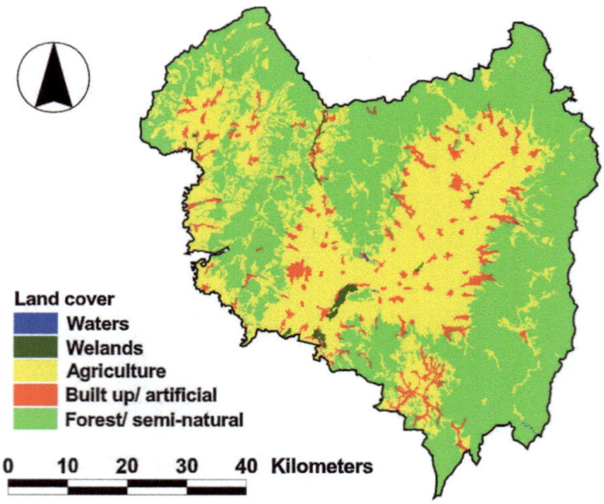

Fig. 9 Land cover in Covasna County

In the following paragraphs, the potential effects of land cover and use are assessed against the assets of Covasna County. First of all, Fig. 12 presents the natural protected areas. Most of them are of interest for the European Community, as they were declared under the Natura 2000 program as Sites of Community Importance (SCI) and Special Protection Areas (SPA). Parks (national and natural) and reserves (including scientific and natural reserves, natural monuments) represent a little percentage. Most of these areas overlap and are situated in the northwestern and northeastern parts of the city. The protected areas reflect very well the

Fig. 10 Land use in Covasna County

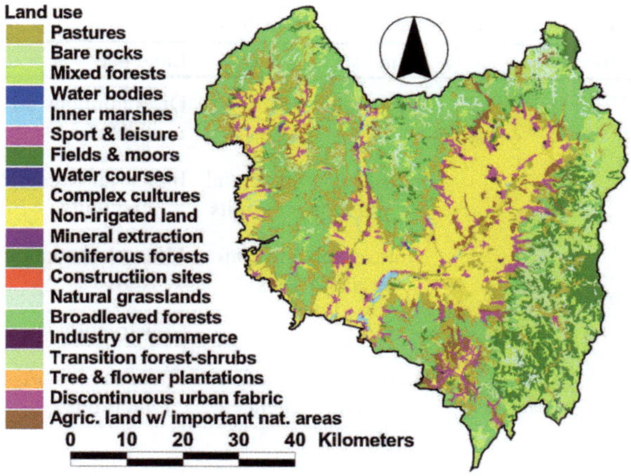

biogeographical (Fig. 13) and ecological (Fig. 14) diversity of the county, and also its geographical geography, as indicated by the elevation (Fig. 15).

The most important pressures against the natural protected areas are represented by envisaged climate change and land cover and use. In fact, the two phenomena, as well as the unwise use of energy, make up together with their relationships what is called global change (Dale et al. 2011). Almost all natural protected areas are likely to be affected by temperature increase (Fig. 16) and most of them also by lower precipitations (Fig. 17). The most important consequences of climate changes affect natural ecosystems directly and indirectly (Blenckner and Chen 2003; Marshall et al. 2008; Thomas 2003). Impacts on agricultural systems include exposure to high temperatures, changes in the regime of precipitations, decrease in nutrients, exposure to fire, increased erosion due to the winds, and the dispersion of agricultural diseases and pests (Secretariat of the Convention on Biological Diversity 2007).

Figure 18 shows clearly that land cover and use changes occurred during 1990–2000 within these areas as well. The most important cause is deforestation, and even though at the time when forests were cut the areas did not benefit upon the protection status, currently, they suffer from the previous damage, which impairs them structurally and functionally (including ecosystem services). The main cause is the change of ownership from the state to people who reclaimed their property (Roman 2009).

Both observations are relevant for planning purposes. The administrators of the area will have to consider not only enforcing the protection status, but also should take into account ecological restoration, in order to account for the effects of land cover and use changes, and mitigation actions to tackle envisaged climate changes. Also, the contribution of previous forest cuts must be taken into consideration by planners, particularly the massif loss not compensated by natural regeneration or plantations.

Table 2 Land cover and use in Covasna County

Land cover	Land use	Land use	Area (ha)	Percentage (%)
Artificial surfaces (4.40 %)	Urban fabric	Discontinuous urban fabric	14,805.13	4.00
	Industrial, commercial and transport units	Industrial or commercial units	873.67	0.24
	Mine, dump and construction sites	Mineral extraction sites	419.33	0.11
		Construction sites	42.85	0.01
	Artificial, non-agricultural vegetated areas	Sport and leisure facilities	150.14	0.04
Agricultural areas (40.67 %)	Arable land	Non-irrigated arable land	78,951.93	21.33
	Permanent crops	Fruit trees and berry plantations	484.62	0.13
	Pastures	Pastures	51,266.76	13.85
	Heterogeneous agricultural areas	Complex cultivation patterns	8,523.93	2.30
		Land principally occupied by agriculture, with significant areas of natural vegetation	11,322.16	3.06
Forest and semi-natural areas (54.40 %)	Forests	Broad-leaved forest	10,4357.19	28.20
		Coniferous forest	32,868.39	8.88
		Mixed forest	41,065.46	11.10
	Scrub and/or herbaceous vegetation associations	Natural grasslands	5,154.80	1.39
		Moors and heathland	88.30	0.02
		Transitional woodland shrub	17,723.35	4.79
	Open spaces with little or no vegetation	Bare rocks	71.14	0.02
Wetlands (0.44 %)	Inland wetlands	Inland marshes	1,629.76	0.44
Water bodies (0.09 %)	Inland waters	Water courses	167.72	0.05
		Water bodies	130.91	0.04

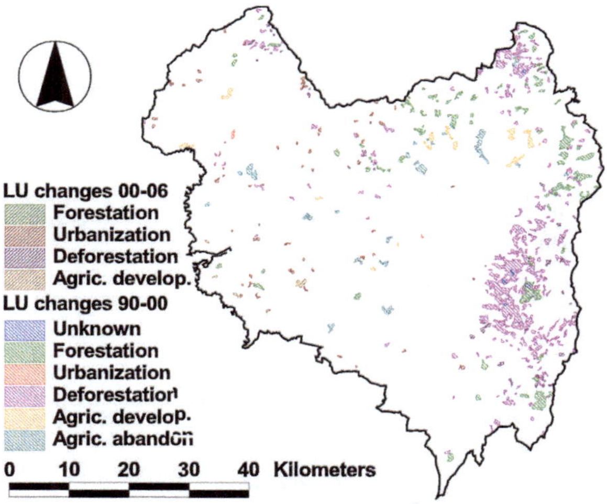

Fig. 11 Land cover and use changes in Covasna County

Fig. 12 Natural protected areas of national importance in Covasna County

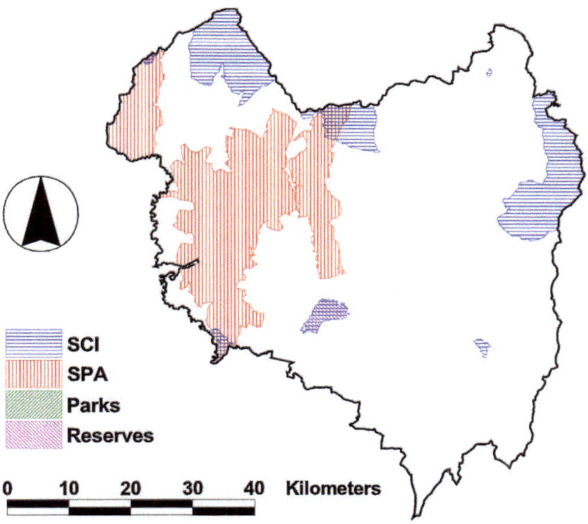

Table 3 Main causes of land cover/use changes in Covasna County

Cause	1990–2000		2000–2006	
	Area (ha)	Percentage (%)	Area (ha)	Percentage (%)
Abandon of agriculture	1,508.06	9.12	–	–
Development of agriculture	1,077.84	6.52	177.99	10.45
Deforestation	9,359.46	56.62	685.79	40.28
Forestation	4,406.34	26.65	19.82	1.16
Urbanization	59.03	0.36	818.95	48.10
Unknown	120.41	0.73	–	–

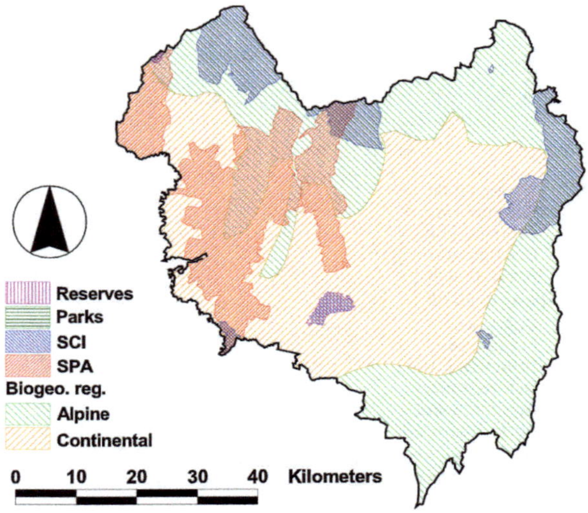

Fig. 13 Biogeographical coverage of natural protected areas of national importance in Covasna County

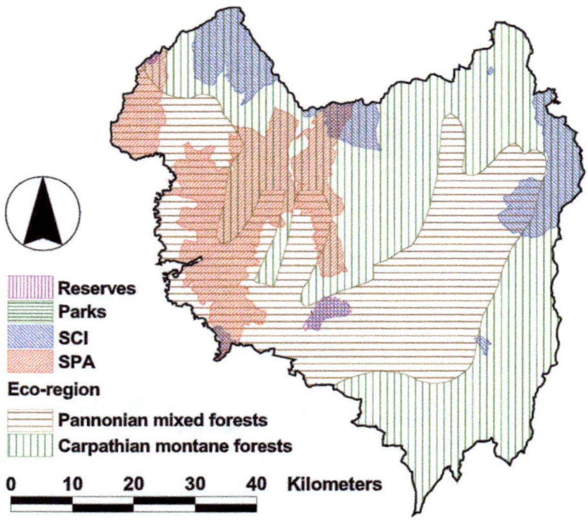

Fig. 14 Ecological region coverage of natural protected areas of national importance in Covasna County

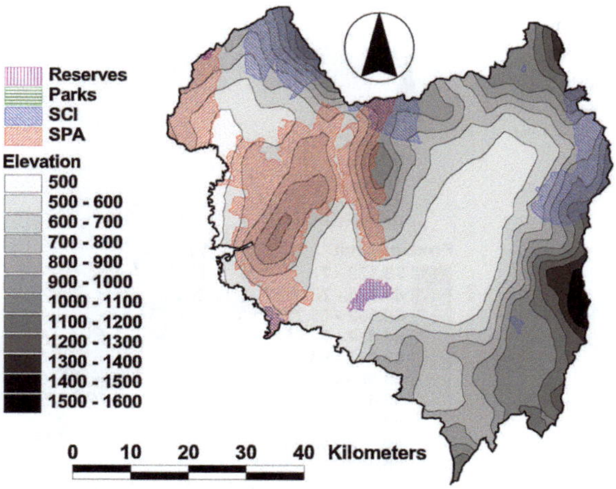

Fig. 15 Elevation of natural protected areas of national importance in Covasna County

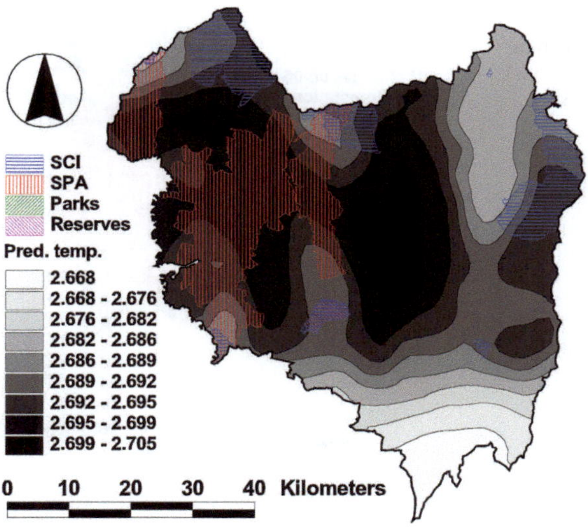

Fig. 16 Influence of high temperatures on natural protected areas of national importance in Covasna County

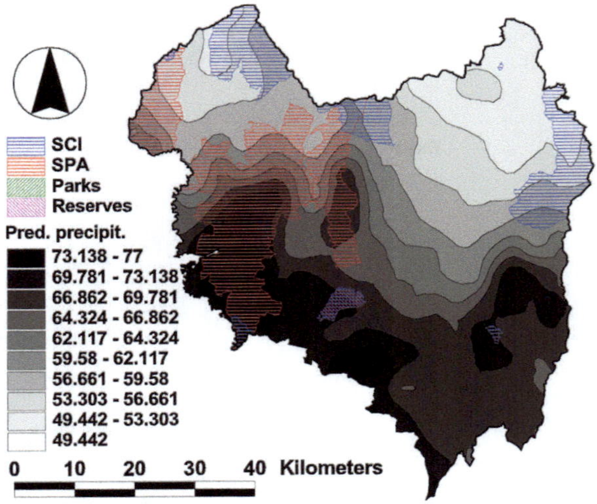

Fig. 17 Influence of low precipitations on natural protected areas of national importance in Covasna County

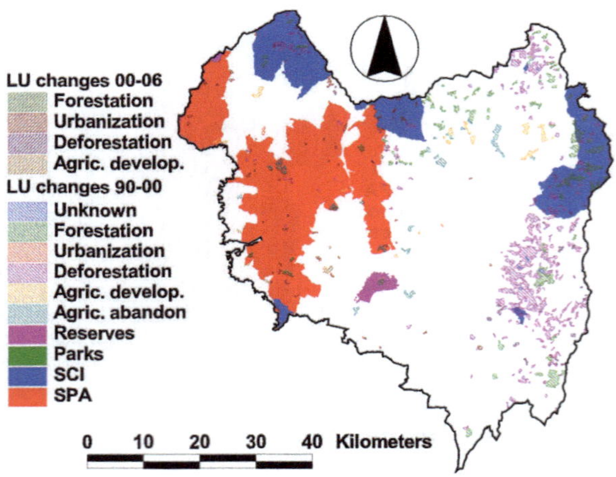

Fig. 18 Land cover and use changes within the natural protected areas of national importance in Covasna County

3 Conclusions

The case study presented above illustrated that the use of CORINE data can be at the core of a desk study looking at the evolution of a Romanian County in order to establish planning priorities. Combining information on land cover and use to the existing situation, particularly natural protected areas, and climate predictions allowed for pinpointing some important changes. These include, first of all, the massif deforestations occurred during 1990–2000, and continuing during 2000–2006, not followed by re-plantation, and second the potential effect of climate change (higher temperatures and lesser precipitations) on the natural protected areas. The use of such data is especially useful provided the lack of centralized data the environmental status of the region.

Therefore, the most important conservation priorities are represented by the forests covering the mountain areas. They need to be integrated into natural protected areas, and their protection status must be enforced. Actions are also required to tackle with the possible effects of climate changes. One possible way is to limit the change of land use.

References

Andrewartha H, Birch C (1954) The distribution and abundance of animals. University of Chicago Press, Chicago

Basiago AD (1999) Economic, social, and environmental sustainability in development theory and urban planning practice. Environmentalist 19:145–161

Blenckner T, Chen D (2003) Comparison of the impact of regional and North Atlantic atmospheric circulation on an aquatic ecosystem. Clim Res 23:131–136

Brundtland GH (1987) Our common future. Oxford University Press, Oxford

Collignon P (2009) Situation et défis du monde rural—Le patrimoine et le paysage au cœur des enjeux du développement territorial durable [Status and challenges of the rural world-heritage and landscape as core goals of sustainable territorial development]. Aménagement du territoire européen et paysage 88:99–103

Council of Europe—COE (2005) Council of Europe Framework Convention on the Value of Cultural Heritage for Society. Council of Europe Treaty Series 199:3

Cowen DJ (1988) GIS versus CAD versus DBMS: what are the differences? Photogram Eng Remote Sens 54:1551–1555

Crăciun C (2010) *Peisagistică* [*Landscape planning*]. "Ion Mincu" University of Architecture and Urbanism, Bucharest

Dale VH, Efroymson RA, Kline KL (2011) The land use–climate change–energy nexus. Landscape Ecol 26:755–773

Dutcă I, Abrudan IV (2010) Estimation of forest land-cover change in Romania, between 1990 and 2006, bulletin of the Transylvania University of Braşov Series II: forestry, Wood Industry, and Agricultural Food Engineering 52:33–36

Gibson RB (2006) Beyond the pillars: sustainability assessment as a framework for effective integration of social, economic and ecological considerations in significant decision-making. J Environ Assess Policy Manage 8(3):259–280

Govindasamy B, Duffy PB, Coquard J (2003) High-resolution simulations of global climate, Part 2: Effects of increased greenhouse cases. Clim Dyn 21:391–404

Haeckel E (1866) Generelle morphologie der organismen. Georg Reimer, Berlin

Hawkes J (2001) The fourth pillar of sustainability: culture's essential role in public planning. Melbourne, Australia

Hijmans RJ, Guarino L, Cruz M, Rojas E (2001) Computer tools for spatial analysis of plant genetic resources data: 1. DIVA-GIS. Plant Genet Resour Newsl 127:15–19

Hijmans RJ, Cameron SE, Parra JL, Jones PG, Jarvis A (2005) Very high resolution interpolated climate surfaces for global land areas. Int J Climatol 25:1965–1978

International Union for the Conservation of Nature and Natural Resources—IUCN (1994) Guidelines for protected areas management categories. IUCN, Switzerland and Cambridge

Jančič M (2007) *Valorizarea peisajului si patrimoniului prin demersuri teritoriale de dezvoltare* [*Enhancing landscape and heritage through spatial development startegies*]. Sixth Meeting of the Workshops for the Implementation of *European Landscape Convention*, Council of Europe, September 21, 2007, Sibiu, Romania

Jensen JR (2000) Remote sensing of the environment. An earth resource perspective. Prentice Hall, New Jersey

Littig B, Grießler E (2005) Social sustainability: a catchword between political pragmatism and social theory. Int J Sustain Dev 8(1–2):65–79

Marshall JD, Blair JM, Peters DPC, Okin G, Rango A, Williams M (2008) Predicting and understanding ecosystem responses to climate change at continental scales. Front Ecol Environ 6(5):273–280

Murphy K (2012) The social pillar of sustainable development: a literature review and framework for policy analysis. Sustain Sci Pract Policy 8(1):15–29

Parliament of Romania (2002) Lege nr. 451 din 8 iulie 2002 pentru ratificarea Convenţiei europene a peisajului, adoptată la Florenţa la 20 octombrie 2000 [Law no. 451 of July 8, 2002 on the ratification of the European Landscape Convention, adopted in Florence, October 20, 2000]. Official Gazette 536(1)

Parliament of Romania (2007) Ordonanţa de urgenţă 57/2007 privind regimul ariilor naturale protejate, conservarea habitatelor naturale, a florei şi faunei sălbatice [Urgency Ordinance no. 57 of 2007 on the regime of natural protected areas, conservation of natural habitats, and wild flora and fauna]. Official Gazette 442(1)

Petrişor AI (2008) Levels of biological diversity: a spatial approach to assessment methods. Rom Rev Reg Stud 4(1):41–62

Petrişor AI, Ianoş I, Tălângă C (2010) Land cover and use changes focused on the urbanization processes in Romania. Environ Eng Manage J 9(6):765–771

Philips A (2002) Management guidelines for IUCN category V protected areas: protected landscapes/seascapes. IUCN, Switzerland and Cambridge

Pickett STA, Grove JM (2009) Urban ecosystems: what would Tansley do? Urban Ecosyst 12:1–8

Roman T (2009) The forest of Romania: a social-economic's dramma. Theor Appl Econ 535(6):57–64

Secretariat of the Convention on Biological Diversity (2007) Biodiversity and climate change. International Day for Biological Diversity Booklet. Secretariat of the Convention on Biological Diversity, Montreal, Canada

Tansley AG (1935) The use and abuse of vegetational concepts and terms. Ecology 16:284–307

Thomas C (2003) Climate change and habitat fragmentation, In: Green RE, Harley M, Miles J, Scharlemann J, Watkinson A, Watts O. Global Climate Change and Biodiversity, University of East Anglia, Norwich, pp. 22–23

Troll C (1968) *Landschaftsokologie. Pflanzensoziologie und Landschaftsokologie* [*Landscape ecology. Plant sociology and landscape ecology*]. In: Tiixen R (ed) *Berichte das 1963 Internalen Symposiums der Internationalen Vereinigung fur Vegetationskunde*. Dr W. Junk Publishers, The Hague

United Cities and Local Governments (2010) Culture: fourth pillar of sustainable development. Policy statement. United Cities and Local Governments, Barcelona, Spain

Vădineanu A (1998) *Dezvoltarea durabilă, Vol. I. Bazele teoretice ale dezvoltării durabile* [*Sustainable development. Vol. I. Theoretical background of sustainable development*]. University of Bucharest Press, Bucharest

Vădineanu A (2004) *Managementul dezvoltării: o abordare ecosistemică* [*Management of development. An eco-systemic approach*]. Ars Docendi Press, Bucharest

Wu J, Hobbs RJ (2007) Key topics in landscape ecology. Cambridge University Press, Cambridge

Part V
Transition Landscape: Periphery/Consumption/Decision—Reappraisal

Urban Culture, Urban Cultural Landscape

Angelica Stan

Abstract The purpose of this chapter is to investigate the relationship between the concepts of "culture" and "cultural landscape," both regarding the actual urban space. As the semantic spheres of these terms overlap, there are many other almost similar approaches, but tributary to specific areas—as "cultural geography," "cultural anthropology," "cultural history," etc.), clarifying and discussing the nuances that distinguish these terms and concepts, customizing them for Romania's situation, become a necessary and useful action. In terms of the architect-urban planner, urban culture is like an umbrella under which takes place the theoretical and practical planning work; finding the common denominator between the various theories that govern this area of research, may seem an utopist attempt. This approach is a sketch of a possible research route in order to identify the core curriculum in the high ramification of studies on urban culture and urban cultural landscape.

Keywords Urban culture • Urban cultural landscape • Elite • Periphery • Center • Political • Vernacular • Coexistence • Multilayered • Tectonic structure • Fuzzy • Connections • Language • Experience • Poetry • Indeterminacy • Vague • Absence • Value

1 Introduction

In the current understanding, the urban culture refers to official culture, somehow synonymous with center one or the elite one, which belongs to the intellectuals, journals and "cultured" people. It seems to contain a set of elements which forms

A. Stan (✉)
Faculty of Urban Planning, Department of Urban and Landscape Design, Ion Mincu University of Architecture and Urbanism, Academiei 18-20, 010014 Bucharest, Romania
e-mail: angelicastan@yahoo.com

C. Crăciun and M. Bostenaru Dan (eds.), *Planning and Designing Sustainable and Resilient Landscapes*, Springer Geography, DOI: 10.1007/978-94-017-8536-5_8,

its profile—from history, memory of the places, personalities, literature, art, to architectural style, way of using the public spaces, way of "consuming" the city, entertainment, music, clothing, etc. At the same time, through the connection that it has with the generic model of modernity, the urban culture participates in shaping the identity of the urban space and mutually, the specific and qualities of the urban space gives to urban culture its own personality.

Thus, the urban culture is accepted as a stance of the whole culture of a nation, an expression of its performance regarding the affirmation of its own identity; but, in this way, many other expressions of national culture—as it is rural culture, or the culture of marginal groups, for example, are lost.

A key area of debate concerning the urban culture is the relationship between the dominant groups ("elite") and the subordinate ones ("popular") (Burgess and Gold 1985). The tension and the significance of each are expressed in the "political" and "vernacular" landscapes (Cosgrove and Jackson 1987) or in cultural differences between central place and periphery (Stan 2009).

2 An Integrative Approach to Urban Culture

Nowadays, in the cities culture study area, there are many developments which provide nontraditional ways of "reading" cities. Current understanding of urban culture is somewhat burdened by increasingly narrow angles in which it is seen, but urban cultural landscape cannot be restricted to a specific result or product. Returning to an overview of urban culture as a fact intimately connected with man and human civilization, cleaned of any intellectual vanity of multiple "culturalist" disciplines, could be a solution. The position of this paper is situated in a middle place between an absolute humanistic approach of culture and a scientific research about cultural landscape, taking into account that landscape itself it is both poetry and mathematics, esthetic and science, imagination and experience (Stan 2011).

We recall that the very concept of *cultural landscape* comes to overcome the old and ineffective opposition between nature and culture, having thus an anthropological foundation: culture *is* "human nature," and all people have the capacity to symbolically encode their life experiences in their language.

The landscape approach of urban culture tries to affirm the fact that language is a modality of being-in-the-world, part of our experience in the urban space (Ricoeur 1976), so "language" and "experience" (and the subsequent tension relationship between semiotic and phenomenology) are part of the same sphere of interests, which needs a holistic way of thinking.

So, we are situating far from the old definition of cultural landscape which says: *the cultural landscape is fashioned from a natural landscape by a cultural group. Culture is the agent, the natural area is the medium, the cultural landscape is the result* (Sauer 1925). More appropriate for us seems to be the earliest articulations of the anthropological meaning of the term "culture," coming from Sir Edward Tylor, who writes on the first page of his 1897 book: *Culture, or civilization, taken in its broad, ethnographic sense, is that complex whole which includes knowledge, belief, art, morals, law, custom, and any other capabilities and habits acquired by man as a member of society* (Tylor 1920).

3 Urban Culture in Romanian Space

When referring to Romanian space, the history of the culture is a mixed one, in a perpetual inadequacy and inability to join a clear pattern, always swinging between Occidentalism and autochthonism (Alecu Russo vs. Vasile Alecsandri), or between criticism and synchronicity (Titu Maiorescu vs. Eugen Lovinescu), or between Edgar Papu's protochronism and Constantin Noica pattern of "resistance through culture" (Neagoe 2008). All these pairs of opposites were active and still operate in a harmful manner for Romanian culture; in these confrontations, usually win the third-one, the nonrepresentative, the minority, the marginal, the worthless, one who, taking advantage of the confusion, manages to impose itself as dominant model. On the other hand, the Romanian urban culture is a relatively new phenomenon, not older than two centuries—the cultural modernization is still in progress. Its main feature that remains crossing over all stages and evolutionary patterns is still derived from this permanent need for connection to modern values, either by imitation, criticism, adaptation, or transfiguration.

In intellectual's speeches, two extremes are met often, yesterday and today: On one hand, there is a noncritical mimicry which imports indiscriminately any form of Western culture, on the other hand, in the opposite direction, a stubborn refusal of modernity and an exaggeration of the native/past values. Let us not forget that urbanism was consolidated as discipline only in the last century, by the transition from the quasispontaneous practice to build, to urbanism as planning system and doctrine, focused on the quality of the urban life.

And if in its whole, the Romanian culture owes the essence to literature and literary criticism efforts, the urban culture begins to be indentified here only in the interwar period, with the affirmation of the modern architecture through the creations of Marcel Iancu, Horia Creanga, Ion Mincu, Cincinat Sfintescu, etc. The intellectual effort of interwar period has issued Romanian urban culture of rural models and of limited understanding as a local, parochial culture.

From the early twentieth century, the Romanian urban culture was one in which official architecture was the main pawn, maintaining the connection between elite and Western modernity. At the same time, the concrete reality of the urban space offered a contradictory landscape: the coexistence of luxury and misery, a mix of the European way of life with the Eastern one, of the provincialism and the eternal inferiority complexes—a cultural landscape of the uncertain artistic, hybridized taste. The Romanian urban culture was not then fully consolidated, the practice of the urban space did not validate the forms and architectural concepts, and the street culture was reduced to a series of social conventions and behavioral habits.

But, understood as a living phenomenon, the urban culture is closely linked to a specific lifestyle of the urban population, in relation with the type of group/community to which it belongs, correlated with an attitude or personal expression, but also with the urban space that hosts them. It is closer to the concept of *habitus*—as nonconscious system of states derived from the economic, cultural, and symbolic capital of the subject (Leach 2006).

Urban culture is one of the actual expressions, full of so many meanings, which it cannot be defined directly anymore, but it requires subsequent clarifications, from various points of view. The anthropology supports the definition of urban culture as a creation of a stable human settlement in time and space (Lawrence-Zúñiga 2003), an emanation of the sedentary human being who develops a sense of belonging to a specific place, illustrated through the way in which he cares and "cultivates" the spaces (as opposed to the nomad's, less tied to a certain territory to which he can impregnate his own values). The ancestral sedentary of the urban population could be, perhaps, the source of his ongoing need to define his imprint on the urban space, to continuously model and improve it as a vital space, to creatively assimilate the influences, to create representativeness, quality and adding value-things. Each animal species creates its own perfect shelter and builds its own perfect environment, this being linked to survival instinct (Appleton 1996). In a more sophisticated way, but having as base the same instinct of security, survival and knowledge, the man, both at an individual and social level marks, builds, cultivates, invests economically, psychoemotional, and social the space he lives in, the urban culture being the synthesis of these actions. To dwell becomes synonymous with being (Heidegger tells us), an expression of the way in which the cultivation of the space serves to profane or spiritual purposes.

4 Urban Culture as a Multilayered Concept

Nowadays, the cultural meaning of "urban" is related to the need of a certain comfort required from the urban space. Beyond the spatial-functional insurance of this comfort, there is a need of experiment, of exploration, of cumulative (and somehow undifferentiated) information—a comfort of freedom, maybe. We consume what the city offers to us, without permanently paying attention to the content of this information, sometimes harmful. We consume the city and it consumes us, in its turn. The information that the citizen "asks" for is not a necessary one anymore, but one that comes from the exacerbation of an instinct of information: We are "programmed to explore" from the beginning, gathering the information as a germinating ground for future creations (Appleton 1996). The creation of the urban culture was achieved, over centuries, thanks to this instinct of knowledge through personal or collective experiences, through gestures of acceptance or denial, in a climate of freedom of choice and expression.

A necessary multiple and open lecture requires to see the urban culture as a *multilayered concept*: First is a nucleus layer which consists of everything that was formed through crystallization in time, in an "official" way (and which tries to keep representative), equivalent to its cultural heritage—material or immaterial: history, literature, art, architecture, places, personalities, habits, traditions, etc., widely known and entered in the public consciousness, often used in tourism, in the international promotion of the city, and on which the city can affirm its competitiveness. These things come to be understood, though, in a stereotyped way, resulting in prejudice, and often altering the perception of the real urban culture, the one that essentially forms the second layer of this concept.

This one has a less defined structure, a "fuzzy" or liquid layer due to the everyday life, and it is composed by different elements—from the dominant type of the physical atmosphere, the people's usual behavior, to the type of language, the urban legends and unknown personalities from clubs, bars and blogs, from the "streets' children," to poetry and live music, from the way of public advertising, to the periphery's poverty, aggression, fraud, violence, etc. If the first layer of the urban culture, present in the school books and travel guides, is a quasipermanently visible one and recognized as such (slightly ever disputed), this secondary layer is visible at every moment in the individual forefront, but it is less recognized, somehow seen as insignificant, ignored, or suppressed by the pressure of other constraints. It is also the layer of personal experience of the urban life, of the daily contact with the city, which impregnates everyone's imaginary, which is consumed in banal facts and actions/interactions, which implies a certain routine in assuming the condition of the everyday life and which, not least, is expressed in various ways: immediately suppresses in street language, in street art, in the way of looking at each other, it creates reactions in the media, limit attitudes, or, on the contrary, it is camouflaged, to be revealed later, in literature, poetry, underground art. The diversity and heterogeneity arising from this wave of individual or quasicollective expressions are not capable of counteracting the "nucleus" zone of the urban culture phenomenon, but it is like a ferment of its evolution, as a source of growth and development, in a spiritual, social, and economic meaning.

Besides all these, we can find a third "layer" of this "urban culture" theoretical construct: that of the fiction, of disinformation, of deconstruction. It is like a metatext carried by each component of urban culture, spontaneous, free of references and fleeting. A layer of *all connections* that largely occurs in cyberspace, but also of those from social manipulation level, from all induced states, ambiance, fashion, rumors and that come together to reinvest the real spaces, places, language, taste, creations and everyone's opinions with some parasitic or contradictory elements, creating flows, waves, ridges visible in the *magma* of urban knowledge and experimentation. Thus, the urban culture would be seen as a *tectonic movement*, with a "superficial motion" from which, at long intervals of time, some parts pass to the stable layer and become classic, petrified, and others burst into everyday life—as "geysers" of the cultural events, civic attitudes, or individual gestures, that sometimes succeed to catch our attention (Fig. 1).

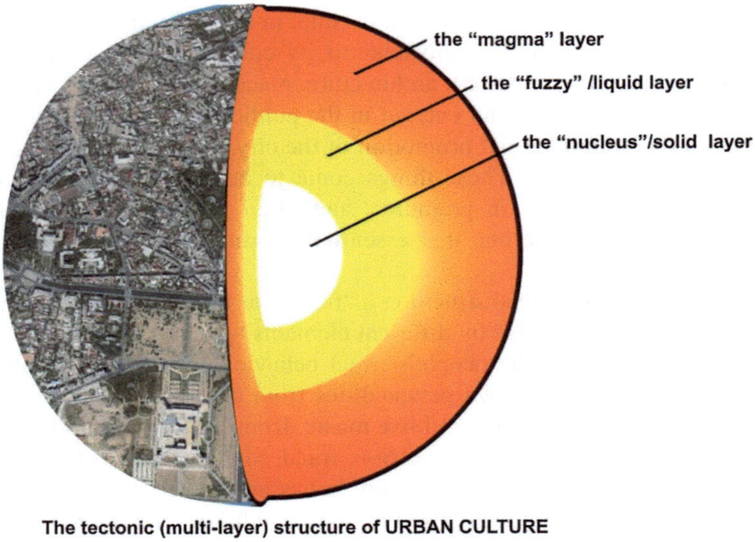

Fig. 1 The tectonic (multilayered) structure of urban culture reiterates the Earth's structure. What we see and live and call as city, is the very skin of an entire deep "geological" process; *Source* author

5 Urban Culture as Poetry

The more these latter two segments/layers are active, the more the poetic connection introduces new instruments of city experimentation for each of its dwellers, consumers, creators or destroyers, the more the chance of culture's nucleus to live increases, its incidence on the man deepens—and this already means to grow sustainably, to grow alive, keeping values for future generations.

There are already very pertinent opinions according to which the poetic, by its subtle, but effective methods (not in writing, but in thinking) have a certain responsibility in shaping the urban culture, from the interior (which means at an individual level), introducing a vector of additional and alternative knowledge, a sensitive discourse both intimate and social, a psychopoetic mechanism of resistance to everyday misery (Farley and Roberts 2009). And may be the fact that today's urban culture in Romania is so precarious, so impregnated by stupidity-demagogy-hypocrisy-kitsch is not only because of the politicians, media, economy, mentalities and unfortunate history of this people, but rather because of a lack of "poetic discipline", properly admitted (and not rejected) within each inhabitant of this country. Once this instrument of interior validation of the exterior world is compromised (from school even, since it is trying to annihilate any capacity to perceive *poetry*), there is a dilution of this once healthy resort of evaluating things, which starts from sensations, moods and then it allows interpretations and personal judgments (Fig. 2).

Fig. 2 Inside the urban culture, there is a need for poetry; collage, *Source* author

The *urban culture as poetry* (discourse, but also action and being—in Heidegger's terms) is an issue that deserves a special attention, especially seeing how in Romania, now, the cultural reaction to the "crisis landscape" is one of retraction in a subversion zone. The resistance is consumed by individual acts, atomized, unable to provide communities, as long as the urban public space is invaded by political, demagogic, and aggressive advertising gestures.

Looking at the things from a different angle, the urban culture can be understood as an expression of emancipation comparing with the village world, or, equally, as assertion of power and importance of the center compared with the periphery. In everyday language, the term "urban" seems to refer to an intrinsic quality, under which the defects, inadequacies, the lack of measure would be less visible. Always and everywhere, the urban culture also means legends, images, and urban myths born beyond the *mainstream*, resulted from assuming the liberty to remain anonymous, even from the pleasure of being camouflaged in the crowd. On the other hand, *if I live in the city it doesn't mean that I am part of the urban environment* is one of the opinions, apparently oxymoronic, relatively recently appeared in the literature about "urban." The urban culture is perceived as an environment that fosters and promotes certain behavior, namely urban, but the rules of this behavior/style become less and less clear. It is interesting that, with this assertion, the urban culture seems to not be strictly linked to the city as a place. Consequently, *you can be urban even if you live in the countryside*: The *urban* becomes an attribute that must be wholly recognized and validated by a certain community/group (involving the surface layers of the "urban culture" concept, denominated by behavior, opinion, language, clothing, etc.).

We could accept this pattern of the urban culture *separated* from its physical base, if we could believe that between the man and the urban space itself there is only a superficial relationship, strictly "utilitarian". But no matter how sophisticated and efficient, no matter how powerful the universality and globalization are, the virtual space fails in replacing the physical space, imprinted by man territory, its vital matrix. The need of urban space, beyond the biological need, is both an esthetic and scientific one, tied to the necessity to check and compare, in particular, in situ, the character and appearance of the cultural patters conceived/viewed in abstract.

6 Urban/Suburban Culture and Landscapes

Faced with the crisis of its own persuasion tools, the *urban culture* cannot validate itself in other way than by reference to something considered as an opposite term: In this case, the *suburban culture*, closely related with the "concrete space," expressed by the direct perception of a diversity of situations, characters and landscapes, socially, ethnically and esthetically charged. The "suburban" culture wrongly considered a secondary and unfortunate type of urban creation; it is a world in itself, too complex to be sent as a failed product of the city. The suburban culture must be understood as part of the urban culture, perhaps the raw material from which the latter is built. The city itself, at the age of congestion, segregation, and vulnerabilities, requires a different understanding of the suburban culture: as a sharp reply of its old history, as a protest from interior, as a way to deny the urban mechanism that designs and kills, with the same fervor, its own values.

As a wide phenomenon, the urban creates its own landscape, and this is a cultural one, as it is a faithful mirror of the whole composition of patterns and values of a specific society. The cultural landscape linked to the urban environment can have notable differences depending on the typology of the physical framework which it refers to; it is a palimpsest and at the same time a kaleidoscope of styles, fashions, manners in which there are mixed, in varying proportions and positions. A common stereotype of thinking is to believe that the spatial and social hierarchy of the city reiterates itself in a similar cultural hierarchy, keeping the positive valence for the center/elite groups and the negative one for the periphery/marginalized groups.

On the contrary, the cultural landscape of the suburban space (from closer periphery to satellite-suburbs) is a distinct type of landscape, resulted from an environment found under the same fate as the city, paradoxically extending its profile, violently undermining its existence, on one hand, and on the other hand, offering an alternative. The cultural landscape of urban periphery is a *landscape in transition,* a contradictory resource, corresponding to a process of change of the reference systems at the level of the whole urban system, including indeterminacy, chaos, the vague, incoherency, and absence (Stan 2009). Despite its often desolate appearance, esthetically classified as negative, the cultural landscape of the margins is one loaded with authenticity and potential, being able to offer the urban culture a consistent field of regeneration.

7 Conclusion

Between tradition and modernity, between heritage and innovation, center and periphery, elite and marginal, the urban culture autodefines itself continuously, in real time, in the physical space of the city, and also in its virtual space. When the urban culture is a too loose structure, extremes mix indiscriminately in there, and the dominant is represented by stereotypes, simulacra, insolence and even violence, inconsistency pump, excesses, ambiguities and the eternal *form without substance*.

For a stronger urban culture structure, entire life and city management should lead to seriously taking into account the individual and social experience of urban space. Central urban culture and the suburban culture should not be considered as polarized position, but as two complementary elements that form the culture of this evolutionary, complex, ambiguous and in permanent change entity, which is the actual city. The urban cultural landscape should be regarded as such an *open cultural work*, a continuous creation of social, economical and political forces, of professionals, but also of the inhabitants themselves and especially of time—past, present, and future. The urban cultural landscape is this *common creation* to which the creator, as the consumer, participates equally, and the product is continuously premastering, because *any consumption, is an interpretation and an execution, for in any consumption, the work lives in an original perspective* (Eco 2002).

References

Appleton J (1996) The experience of landscape. Wiley, New York

Burgess AJ, Gold RJ (1985) Geography, the media and popular culture. Taylor & Francis, London

Cosgrove D, Jackson P (1987) New directions in cultural geography. Area 19(2):95–101

Eco U (2002) Opera deschisa. Ed. Paralela 45, Ed.II, Pitesti

Farley P, Roberts MS (2009) Edgelands: journeys into England's true wilderness. Johnatan Cape, London

Lawrence-Zúñiga D (2003) The anthropology of space and place: locating culture. Wiley, New York

Leach N (2006) Uitati-l pe Heidegger (Forget Heidegger). Paideia

Neagoe L (2008) Depăşind modernitatea: crize şi modele în cultura română. Cuvantul 1–2:12–13

Ricoeur P (1976) Interpretation theory: discourse and the surplus of meaning. TCU Press

Sauer C (1925) The Morphology of Landscape. Geography 22:19–53. University of California Publications

Stan A (2009) Peisajul periferiilor urbane. Revitalizarea peisagera a zonelor periferice. UAUIM, Bucharest

Stan A (2011) Poezia orasului. Urbanismul (serie noua) (8):31

Tylor E (1920) [1871]. Primitive culture, vol 1. J.P. Putnam's Sons, New York

The Landscape of the Urban Peripheries: An Alternative Therapy

Angelica Stan

Abstract The landscape approach of peripheries moves the accent on the need to understand the general health of the city and to treat its peripheral territories through an alternative method then the classic regulatory planning methods. The chapter argues by appealing to principles of homeopathic medicine, advancing the concept of "urbanopatie," but not before underscores the importance of exploring and recognizing the suffering of peripheral landscape, as a pre-diagnosis stage. Landscape urban peripheries are crystallized by synthesis of public needs, individual initiatives, everyday practices, legal/illegal effects (direct or indirect), aspirations and cultural background. But this landscape is first an atmosphere that impresses negative, an amalgam that makes it being the source of a strong ambiguity. Re-learning to see these ambiguous marginal urban territories is a necessary step for understanding them, able to highlight their vulnerability, which is not only theirs, but of the whole city.

Keywords Periphery • Alternative • Therapy • Urbanopatie • Transitional landscape • Disease

1 Introduction

The peripheries of the city are areas over which the urbanism acts based on an urban project, committed to a political program and a certain social order or request. The urban regulation, the operation and implementation of an urban

A. Stan (✉)
Faculty of Urban Planning, Department of Urban and Landscape Design,
Ion Mincu University of Architecture and Urbanism, Academiei 18-20,
010014 Bucharest, Romania
e-mail: angelicastan@yahoo.com

C. Crăciun and M. Bostenaru Dan (eds.), *Planning and Designing Sustainable and Resilient Landscapes*, Springer Geography, DOI: 10.1007/978-94-017-8536-5_9,

strategy lead to the application of some principles of urban development with a high degree of generalization to some quite different urban situations. So, it is the "case" of the peripheral areas, in constant motion, entailing particularities that deeply distinguish them from the rest of the city and over which the regulatory urban planning cannot longer act beneficial, as long as it does not take into account, from the beginning, their specific condition within the urban system.

In general, the urban research has focused especially on center, as main depository of urban significance, or, at best, on the city in its whole—considered to have a relatively homogenous structure. The issue of the urban peripheries, somewhat quietly accepted as something different within the general theory of the city, is more and more found in the regionalist discourses and the ones referring to the metropolitan areas, defending, quite often, the coincidence and even the confusion between *periphery*, *suburban areas*, *proximity areas* and *the metropolitan area*, all having in common the issue of the city growth and its impact on surrounding territories. The distinction in the case of the urban periphery—especially that of the big cities—comes, on one hand, from its morphological and spatial–functional complexity (including fragments of urban tissues profoundly different, rural fragments, luxury residences and disadvantaged areas, large industrial, infrastructural implants, etc.) and, on the other hand, from its position of outpost of the city growth. Also, representing the city youngest organs, new centralities emerge with the decentralization and deconcentration in relation with the "mother-city," to which the peripheries manifest a relative dependence. We may say that in terms of functional and economic empowerment, the peripheries are early stages of suburbanization. Their landscape should be understood from the same perspective, as a *transitional landscape*, from the typical urban form to another one (not necessarily more advanced).

Moreover, the urban periphery could be understood as a place of manifestation of the urban growth impact on human life and landscape. It represents a moment in which the urban system equilibrium is re-discussed on its evolution scale, a moment of a possible breaking of the balance between the anthropogenic and natural environment, reminding the urban population that this fact reflects on their own health and the health of the future generations. So, the issues of sustainable development are closely related to the landscape of urban peripheries, which, by its nature, reflects this fragile balance between city intimate resources and global consumption.

2 Exploring the Landscape Peripheries

With the new logic of the cities become metropolitan areas and/or urban agglomerations, in the middle of the IT era, along with the new way of urban space consumption, a new report is created in the center–periphery equation: The two terms are not necessarily antagonistic. New meanings emerge with the re-discussion and revalorization of this report. All over the world, the more or less controlled development of the marginal areas of the big cities is a process with a powerful economic and political engine; the pressure of the built space on the "free" space is

Fig. 1 The "disease" is here the failing to integrate into the landscape a difficult presence as it is the heating element above ground, exposed as an internal organ of the city, in a "dissected" peripheral space in Braila former industrial zone, Romania, 2012. *Source* Author's personal archive

greater as the "power" of this engine is stronger and more unconscious. This pressure produces a different macro-landscape, with unclear features at first glance, in which it is mixed, in different proportions, the natural and the anthropogenic, the urban and the rural and the formal and the informal (Fig. 1).

The first step in learning how to understand this vulnerable landscape of the peripheries is how to look at it. Often, the appearance makes us to take things as they are, but in this case, this attitude is wrong. As a territory very little known and often unobserved (where is not a touristic interest), the landscape of our peripheries must be first of all explored as places of possibility, mystery and poetry (Farley and Roberts 2009).

3 Re-learning to See the Periphery

The new understanding of the periphery focuses on this *in-between territory*—in *and* between urban and non-urban, a "middle landscape" (Rowe 1991), tertiary landscape, a "third state" of the city or "le Tiers paysage" (Clément 2004) overwhelmed by its diffusion. Our present incapacity to cannot see what is *in fact* the peripheries landscape is partially coming from a weak resistance to a very strong movement, The New Urbanism current, which is trying to cope with the urban sprawl phenomenon. In recent decades, this movement has led to the creation of an entire literature oriented to measures and procedures for planning and urban design to counter the unsustainable growth of the city. But notice that, in fact, all these measures try treating symptoms (territorial expansion) and not the "disease" itself (peripheries phenomenon), which is an older disease of cities, never really treated. Another way of looking at landscape peripheries becomes necessary, highlighting this issue as a moment of partial overlapping between the classical *urban order* and the *vague* order of the periphery, as an opportunity of city boundary re-articulation. A sensitive lecture oriented to resurrect the land potentiality past over the common sense: residual, neglected, vacant spaces, spaces pending the clarification of their own legal status, industrial "friches," ghettos, residential areas (new and/or compromised from the design stage), everything that the city refuses/ rejects could be taken and they become "raw material" for a potential landscaping act. The landscape therapy on urban peripheries would act as under a "Hippocratic Oath": trying to cure the sick urban landscape, to heal the city patient, primarily using its own resources (Stan 2009).

Learning to look at the peripheral landscapes without seeing only the unpleasant and the ugliness, the kitsch and the imposture, the fail and the illegal, considering the periphery as a "recoverable organ" of the city, this (relatively) new profession of *urban landscape planner* has the task of trying to restore the equilibrium from a middle position, often ungrateful.

The real implication of this landscape approach in the urban planning field brings not only additional multidisciplinary information, but also exercises a *critical function* in the professional practice of the planners and architects. But this criticism should not be understood in destructive terms, on the contrary. The stake of this alternative approach is to bring to light the "camouflaged" world of urban periphery landscape. Perhaps this is too little for theory and cannot be a symmetrical movement to New Urbanism, but is necessary, as long as it is honest. By opposite, the term of landscaping is understood by some authors as one that *excavate a paradox in the world: disguise the physical facts of raw human activity, cover human tracks, conceal by camouflaging the terrain according to some orderly idea of the natural* (Farley and Roberts 2009:15).

Exploring the peripheries can also mean re-learn the *metaphorical* reading of its components: for example, the *container,* an ultra-present object in the peripheral landscapes, carries with it a whole symbolism and a rich content of ideas. Often denied by architects not being enough stimulative for their work, hated by landscape planners, ignored by urban planners (for the master plans' scale it does not exist), the container became *an essential image of our epoch, a meme* (Farley and Roberts 2009).

The urban therapy through peripheries' landscape calls to what George Simmel once remarked to be the *Stimmung* of the landscape—something between objective atmosphere and mood of the observer. The treatments of these problematic urban areas based on the perceptual unity of the landscape encompass the affective reflex or the social affective capacity. So perhaps the only valid methodological model is one that goes from bottom to top, from local and individual, to the collective/social experience and then, to the administration (Ritzer 2008).

Thus, the abandonment of the "peripheries" could not be judged as only one of the authorities—although it is the easiest to identify and accused of; it is also an abandonment of the inhabitants themselves who do not manifest (anymore) a feeling of belonging to the places where they live in, or which they will relocate in, the periphery being considered a moment of transition, as a temporary moment in their existence. Therefore, the landscape generated through daily or occasional practices becomes one that lacks symbolic and representative baggage, that permanence and continuity of human existence in a given territory makes it possible.

4 The Diseases of Urban Peripheries

This dialog between the city and its own peripheries is welcome and it creates the conditions for a re-settlement of the issue of the *identity* of the city, reduced (in last decades) to a patrimonial vision, overlapped to the issue of the center and its institutionalized values. Also, this dialog can be done in a corollary of the presence (more or less discrete, more or less fixed) of *the city limit*, one that forms a "halo" around all discontinuities of the urban structure, around which is disputed, in fact, the fate of this in-between territories.

Therefore, it becomes interesting to interrogate again on this *limit*—as main tool of the periphery profile: limit in movement, flexible, fragmented, duplicated, diffuse, folded ribbed, line limit, point limit or surface limit, generated by this dynamic of the city toward the neighboring territories. The formal and functional transfiguration of the *limit* make possible the understanding of the periphery as a distinct part of the actual city, but part and entire in the same time, recreating, through a fractal dimension, the city structure itself. Extending the assertion, the disease at *macro-territorial* level falls at the micro-territorial and individually level, affecting the human body *through* the urban body.

Similar to a *membrane* located between the biotic/abiotic and cultural environment, the peripheral surface exposes within another wider surface, (territorial system), enabling and mediating a communication and an exchange: a double mirror of a *state of health* (Fig. 2).

Typical urban diseases are known already: They are human-induced diseases, among other factors (social, biological and psychological) and environmental factors (in this case, urban). The most common are illnesses related or aggravated by all types of pollution that characterize the urban phenomenon. But human diseases in urban environment are just a simple response of diseases at a broader scale, in which man, as individuals and as a species, lives.

Often, and especially in highly industrialized countries, urban periphery diseases are primarily due to poor hygiene conditions that characterize public marginal areas of big crowds. We can distinguish one category of diseases observed on the poor in the industrial outskirts of major cities, due to the exposure to lacks of water, food and access to urban facilities and to modem medical care.

According to other authors, the digestive system and especially one of the intestine, are present not only in peripheral areas, but also throughout the urban area, due to factors resulting from technological innovations, namely, the prolonged exposure to technology-related chemicals (Schedl 1979). Putting this in conjunction with more severe reduction in agricultural areas due to territorial expansion and with the import of food from areas more far away from the city, we can state *that peripheries, canceling their function that were once agricultural, as a buffer between rural and urban, facilitates the spread of many diseases in resident population.*

Another observation arising from the study mentioned above is that the incidence of serious bowel disease (colon cancer) is higher in urban populations in

Fig. 2 The "ghetto" status, sometimes a synonym for periphery, is not just a figure of speech. It is a mental state induced by the proximity sick urban space, whose traits are disruption, misery and abandon. Limit of Militari District in Bucharest, 2010. *Source* Author's personal archive

developing countries and lower in the developed countries. Putting this in context macro-territorial European level, the periphery (this time a politicoadministrative) acts as an environment for development of these urban diseases.

Another category of urban disease observed in peripheral areas is that related to the nervous system. According to some authors, we can speak about *space diseases* that host, facilitate and aggravate the human diseases (as depressive disorder, space anxiety, panic disorder, etc.). *Agoraphobia* is one of them, detected from over 100 years since Westphal identify a group of symptoms experienced by patients walking in streets or open spaces (Westphal 1871). These consist in dizziness, anxiety, palpitations and trembling (Gelder 1982). Later, with the development of cities and territorial expansion, these symptoms were described differently as "fear of space around" (Marks 1981). In other studies, a strong relation was observed between the degree of space heterogeneity in big urban centers and the mortality from ischemic heart disease of elderly (Périssé et al. 2010).

Some other studies reveal a certain relation between mental state anxiety and the living space *scale* perception. Human scale seems to be not just a good condition of spatial organization, a tool for architects, but also a human health condition itself. Space scale is also related to peripheral territory's *fragmentation*. Knowing that, in ecology, any fragmentation of natural habitats brings about a deterioration of living conditions for all biocenosis, the emerging suburbs in natural environments is even more dangerous. Stronger fragmentation in peripheral territory, together with other exogenous factors, became a train of progressive and irreversible distortion of habitats.

Besides this, an unhealthy city has an *urban image* that acts as a declining factor, represented a serious risk to businessmen, farmers and prospective residents

Fig. 3 The ambiguity of this image reveals a peripheral landscape irresolute itself, inducing a state of insecurity by the presence of morphological fractures and a non-urban scales. West of Bucharest, Militari, sector 6, 2006. *Source* author's personal archive

(Goldfield 1973). When epidemics occurred in most major cities in the world, the peripheral areas were most seriously affected by health problems, especially among poorer social groups—the elderly and children. Poverty, lack of easy access to water supply, lack of food and lack of accessibility to health facilities, coupled with isolation and lack of information make that during urban epidemics, these peripheral areas suffer most. Besides this, the peripheries of urban cities today offer a picture dominated by commonplace and standardization, which makes the induced mental state of anomy, carelessness, lack of energy and dilution of the desire to engage in social life. This kind of *image-diseases* soils our look, consciousness and affect, being impossible to feel any aesthetic emotion to any periphery's landscape (Fig. 3).

5 Urban Peripheries Health Assessment: An Alternative Approach Through Landscape

For the situation of urban periphery that concentrates most of the problems, urbanism should be rethought (also) as a science of a *suffering city*, (maybe we can call it: "*urbanopathy*"), trying to understand the urban phenomenon's disease mechanism and all changes caused by it. Perhaps before a diagnosis and before creating urban development strategies, urban planning should (re)become *human*, to relay at the man's heart and mind, including sufferings.

But a proper therapy would require a series of treatments established as a result of many sophisticated and expansive analysis—this being a similar approach to conventional medicine. Another approach could be an alternative one—as the homeopathic medicine is doing and acting totally different from the allopathic medicine. The philosophy of such an *alternative therapy* is based on the idea of a holistic approach of the urban environment. The same as in homeopathy, the patient (the entire city or only its suffering part—the periphery) is treated as a whole system and not by its disease symptoms.

Symmetrical to the homeopathic principle that treats the individual with highly diluted medicinal substances, which triggers the body's natural healing mechanism, that in-turn annihilates the disease condition, the new alternative urban therapy should find specific tools. One of them is landscape—playing here the role of trigger for the qualities of a peripheral space.

As the health is a matter of balance and harmony, landscape mirrors this harmony state of the city. Still relying on the analogy with homeopathic medicine, any landscape intervention in peripheral space can start from some basic principles as:

1. *"Similia Similibus Curantur"*—a therapeutic method of symptom similarity that affirm that "any drug capable of producing morbid symptoms in the healthy will remove similar symptoms occurring as an expression of disease." The principle could became useful in urban and landscape planning—for example, if demolition is perceived as a "morbid" intervention in a healthy urban tissue, it became acceptable when it is about a "sick" tissue, and it produces a better space and life condition.
2. *Simplex*—one simple substance, adapted to each individual; we can treat some "sick" urban spaces re-organizing them totally or just adding a tree; but the solution of intervention should be every time adapted to each situation. Urban regulation should be enough flexible to permit this wide variety of intervention, under a general building rule.
3. *Minimum*—the medicine smallest quantity should be prescribed so as to stimulate the immune system, leading to cure. In some studies, this theory is called the "memory of water": the dilution process, referred to as "potentization," is believed to transmit some form of information/energy from the original substance (Der Marderosian 1996). In landscape planning this principle is correlated with the necessity to provide minimum intervention (in terms of finance involved, but also in terms of disturbance of public and private property. The level of acceptance of any urban or landscape project in a city is related to this disturbance, and people prefer minimum of intervention on a period of time-so the aim is to have project's phasing as fair as possible.
4. *Drug Proving*—a systematic investigation giving the only reliable knowledge that is very essential to cure the disease. Any place in a town—and especially those in periphery—should be considered a "clinic case," and studied "from scratch," observing it in different situation and proving the proposal (by models, 3D simulation, etc.).
5. *Chronic Disease*—a more controversed principle of homeopathy, which says that the first sign of disease is thus always a skin disorder (Campbell 2008). This very interesting principle allows us to think that is always necessary to check out on the urban image of a place, to judge it as a possible indicator of a more profound disease of that zone.

6. *Vital Force*— as human body is governed by a force, a *vital force,* and when this vital force is deranged or weakened, the disease process sets in. As the city is still considered an "live organism," its vital force is necessary to be maintained in an equilibrium—this could be even measured and expressed by indicators as life expectancy, trust in authorities, access to education, pleasure for traveling, etc.
7. *Drug Dynamisation*—the medicinal properties that are latent in natural substances while in their crude state become awakened and developed into activity to an incredible degree. In a complex urban structure, it is important to know how to enhance the city's "natural" defense, as long as it is perceived, by a majority of people, as an autonomous entity. This defense, in peripheral zones, is sometimes the people capacity of transforming places, not in a night, but in decades. This *dynamisation* should be understood in accordance with the space/time scale of each place.

For implementing, these speculative sketches would be necessary to establish more accurate theoretical and practical tools. Moreover, each urban situation in peripheral space requires a *customized investigation* to determine these remedies that should be minimal, properly dosed, noninvasive and stimulating for all biotic, abiotic and cultural systems of any human habitat.

Also, a very important thing in this alternative therapy is *preventing* and involves direct *interaction between patient and therapist.* Thus, the feeling of belonging to the landscape is the one to be encouraged among the periphery's residents and communities, as well as creating an appropriate image, able to signify the new sense of modern urbanity, involving the informational technology in daily life.

6 Conclusions

The alternative approach of suffering peripheral landscape through the concept of "*urbanopathy*" tries to illuminate the possibilities to "treat" in a humanistic way the peripheries, as they concentrate important urban diseases.

Providing a healthy environment for the urban life is not just about those physical, chemical and biological qualities needed, but also providing a *environmental continuum* in ambiance and urban image, so that what we perceive not to cause inconvenience.

This fact, although simple, is quite difficult to be implemented; on the one hand, because it is not brought to public awareness and on the other hand, because people themselves lost the confidence in their possibilities to reinvest emotionally a diseased landscape.

Thus, the recovery (reinvestment, renewal, revitalization, etc.) and treatment of the periphery through landscape "alternative" therapy can create a true opportunity for the entire actual city. This city suffocated by a chaotic development, crushed by formalism and globalization, could be able to re-create its *identity*, assuming its indeterminacy, sufferings and even the errors, as parts of its becoming. Moreover, this alternative approach could be considered as a "prophylactic plan" that would be done *before* the disease become malignant: An urban prophylaxis that could lead to avoiding the future urban mistakes.

References

Campbell A (2008) Homeopathy in perspective: myth and reality. LLU (ed). www.acampbell.org.uk
Clément G (2004) Manifeste du Tiers paysage. http://arlibre.org
Der Marderosian AH (1996) Understanding homeopathy. J Am Pharm Assoc NS36(5):317–321. http://nccam.nih.gov/sites/nccam.nih.gov/files/homeopathy.pdf
Farley P, Roberts MS (2009) Edgelands-jouneys into England's true wilderness. Johnatan Cape, London
Gelder MG (1982) Agoraphobia and space phobia. Br Med J (Clin Res Ed) 284(6309):72. http://www.ncbi.nlm.nih.gov/pmc/articles/PMC1495659/pdf/bmjcred00588-0008.pdf
Goldfield DR (1973) Disease and urban image: yellow fever in Norfolk 1855. Va Cavalc, Autumn 23:34–41
Marks I (1981) Space "phobia": a pseudo-agoraphobic syndrome. J Neurol Neurosurg Psychiatry 44:387–391
Périssé G, Medronho RA, Escosteguy CC (2010) Urban space and mortality from ischemic heart disease in the elderly in Rio de Janeiro. Arq Bras Cardiol 94(4):35–45 São Paulo. http://dx.doi.org/10.1590/S0066-782X2010005000009
Ritzer G (2008) Georg Simmel in sociological theory. McGraw–Hill Companies, New York
Rowe PG (1991) Making a middle landscape. MIT Press, Cambridge
Schedl PH (1979) Intestinal disease and the urban environment. Environ Health Perspect 33:115–126, PMCID: PMC1638099. http://www.ncbi.nlm.nih.gov/pmc/articles/PMC1638099/
Stan A (2009) Peisajul periferiilor urbane. Revitalizarea peisagera a zonelor periferice. Ed.UAUIM, Bucharest
Westphal C (1871) Die agoraphobie: eine neuropathische Erscheinung. Archiv Fiur Psychiatrie und Nervenkrankheiten 3:138

Part VI
Urban Garden: Art/History/Landscape Architecture—Becoming

Viennese Modernism and Landscape Architecture

Iris Meder and Ulrike Krippner

Abstract The text addresses the cooperation of modern architects and garden designers in Vienna in the 1920s and 1930s. Progressive garden architects, among them numerous Jewish women, ran garden design studios with attached nurseries and kept working together with modern architects from the *Österreichischer Werkbund*. Like many architects, most of the very few women working in garden architecture in Vienna in the 1920s and 1930s came from liberal bourgeois Jewish families that had immigrated to Vienna around 1900. Jewish students of architecture usually attended the Vienna *Technische Hochschule*. The *Höhere Gartenbauschule für Frauen* was the first advanced horticulture school for women within the Austro-Hungarian Monarchy and was founded by the women's rights activist Yella Hertzka in 1912; it became the leading training institution for women in horticulture and garden architecture in Vienna and the former Austro-Hungarian Monarchy. This offered women the potential to attain a higher professional training and a basis for economic independence as gardeners or garden architects. Their gardens, together with the architecture of the *Wiener Schule*, were significant contributions to Austrian garden architecture in the early twentieth century.

Keywords Women • Werkbund • Vienna • Exile • Modern architecture • Living garden

1 Training in Architecture in Early Twentieth Century Vienna

At the end of the nineteenth century, a great number of people from the Austro-Hungarian Monarchy immigrated to Vienna, among other reasons, as a

I. Meder (✉) · U. Krippner
Institute of Landscape Architecture, University of Natural Resources and Life Sciences Vienna, Peter-Jordan-Strasse 82, 1190 Wien, Austria
e-mail: imeder@gmx.de; iris.meder@boku.ac.at

U. Krippner
e-mail: ulrike.krippner@boku.ac.at

C. Crăciun and M. Bostenaru Dan (eds.), *Planning and Designing Sustainable and Resilient Landscapes*, Springer Geography, DOI: 10.1007/978-94-017-8536-5_10,

result of the increasing industrialization, as well as the famines and epidemics in Galicia and the agricultural crisis in South Bohemia and Moravia of the early 1870s.[1] In 1910, nearly a quarter of the population of Vienna, which had almost doubled since 1880 and then reached its maximum of around 2.2 million, was from Bohemia or Moravia. Following the emancipation of Jews as part of the Austro-Hungarian Compromise of 1867, a great number of Jewish families, including many from the well-to-do Jewish bourgeoisie, mainly businessmen and entrepreneurs, moved to Vienna (see Wischenbart 1992).

The business of building contractors and the profession of architects did not have a position in the tradition of Jewish consciousness. Over centuries, owning land and real estate was not permitted to Jews of Europe; in the diaspora, it was not possible to feel rooted in an area considered as "homeland." Studying architecture emerged as a considerable theme for Central European Jews only around the turn of the century. At this time, the number of Jewish students in all faculties—among them, architecture—rose dramatically (see Sapák 2000: 14; Meder 2010).

Around the *fin de siècle*, Otto Wagner was the dominant figure in Viennese architecture. His master class at the Academy of Fine Arts, where he taught beginning in 1894, was, however, considered anti-Semitic and therefore was avoided by Jewish students. During his 24 year tenure, there was only one Jewish student in his master class: the architect Ernst Lichtblau.

The leading figure at the Vienna *Technische Hochschule* (Technical University), Vienna's established university for technical subjects, was its president, Carl König. Studying at the renowned Technical University in pursuit of a solid technical education was very popular among Jewish students. Twenty to thirty percent of its students were Jewish, at times even more, whereas only 8–10 % of the Viennese population were Jewish. In the academic year 1909/1910, e.g., there were 2,079 Catholic and 794 Jewish students. König, who himself was of Jewish origin,[2] was a role model for his Jewish students. The art historian and architecture critic Max Eisler spoke of the conservative König, who was also highly esteemed by Adolf Loos, as a "man of tradition [...], who united knowledge and culture"[3] (Eisler 1936: 8). The architect Felix Augenfeld reported, however, that König—in a notice posted on the blackboard—forbade his students to attend Adolf Loos's private building school (see Augenfeld 1981). If this prohibition really existed, many of König's students ignored it. König's historicist approach, which could hardly inspire young architects, together with the rejection of the Wagner School and the aestheticism of the Vienna Secession and *Wiener Werkstätte*, led to the development of a specific *Wiener Schule* of sceptical modernism; its protagonists Oskar Strnad, Josef Frank, and Oskar Wlach, all of them sons of parents who had immigrated to Vienna from the crown lands, worked

[1] This essay followed a two-year research project on landscape architecture in Austria between 1912 and 1945 at the University of Natural Resources and Life Sciences in Vienna, financed by the *Austrian Science Fund* (FWF).

[2] König left the Jewish community in the 1870s and was then without a confession.

[3] "ein Mann der Tradition [...], der Können und Bildung kulturvoll vereinte."

predominantly on housing. In their work together, which began in 1908, they were surrounded by a circle of like-minded fellow students from the liberal, assimilated Jewish bourgeoisie (see Meder 2004). Almost all of them were either born in the crown lands or belonged to the first Vienna-born generation of families that had immigrated to Vienna from Bohemia, Moravia, Slovakia, or Hungary.

In Jewish circles, it was of utmost importance to find a community and both private and professional solidarity in a stable network of family relations and friendships. Josef Frank himself is an example for the importance of the family background in the bourgeois milieu of assimilated Jewry, manifested in the close relationships of siblings who were also successful in their professions: Frank profited intellectually (and probably vice versa) from his contact with the circles of his brother Philipp, who was trained as a physicist and belonged to the *Wiener Kreis* (Vienna Circle) of Moritz Schlick, Otto Neurath, Rudolf Carnap, and Ludwig Wittgenstein (see Meder 2008). The architects' clients were often their friends or relatives and came from the same stratum as the architects; Josef Frank actually worked almost exclusively for friends and relatives (see Meder 2003). The basis of a stable, intellectually open family background was certainly part of the Vienna School's undogmatic approach.

Many architects of the *Wiener Schule* married women with backgrounds in architecture, applied arts, or fine arts and worked together with them on their architectural projects. After 1900, receiving a higher education and professional training had become more important for both men and women. As study and knowledge was important in the Jewish tradition (see Beller 1989), Jewish middle-class families frequently encouraged children of both sexes to attend secondary school (Pass Freidenreich 2002). Hence, 46 % of the girls attending girls' secondary schools in Vienna in 1910 were of Jewish origin (Raggam-Blesch 2008: 50–51; Hecht 2007). For women, in addition to receiving an education in an artistic field, e.g., at the *Kunstschule für Frauen* or the *Kunstgewerbeschule* (see Fliedl 1986), studying at a university was, at least theoretically, also a possibility. Whereas numerous Jewish women were trained as lawyers or doctors, the number of Jewish women architects in Austria was small, albeit relatively larger compared to the number of non-Jewish women architects.[4]

2 Women in Garden Architecture Training

Until the early twentieth century, women in Austria also had no opportunities to obtain advanced training in horticulture and garden architecture. Most garden architects at that time received their training in palace gardens or horticultural

[4] Helene Roth, who studied at the *Technische Hochschule* in Vienna, was the first woman in Austria to receive a degree in engineering. Liane Zimbler, Ella Baumfeld-Briggs, Elsie Lasar, and Ada Gomperz were trained at the *Kunstgewerbeschule*. Friedl Dicker was the only Austrian woman architect who was trained at the Bauhaus.

businesses, and these programs did not accept women. In 1895, the first horticultural high school of the Austro-Hungarian Monarchy opened in Eisgrub (Lednice, now the Czech Republic). To enroll, students had to be at least 15 years old, had to have graduated from secondary school, and had to have served a horticultural apprenticeship for at least one year prior to enrollment (Recht 1976: 16). During a 3 year curriculum, they received training in horticulture, pomiculture, landscape construction, and garden architecture. Well-known Austrian garden architects such as Albert Esch and Titus Wotzy studied in Eisgrub. However, it took another 16 years before the first woman, Luise Waschnitius, graduated in 1911.[5]

At the end of the nineteenth century, middle-class women in England and Germany had opened horticulture schools for women to fight gender discrimination (see Schekahn 2000; Meredith 2003).[6] Because in Austria, there were no opportunities for women to receive professional training in horticulture or in agriculture (Theimer 1909), around the turn of the century "a number of women [...] planned to establish a horticulture school for women and girls of the educated middle class" (Anonymous 1906).

Finally, in 1912, Yella Hertzka (Vienna 1873–Vienna 1948) founded the *Höhere Gartenbauschule für Frauen* in Vienna, the first advanced horticultural school for women in the Austro-Hungarian Monarchy. Hertzka, who was from a liberal Jewish family of the Viennese bourgeoisie, married Emil Hertzka in 1897. During his stint as head of the *Universal Edition*, he turned it into the leading publishing house of contemporary music. Yella Hertzka was intensively engaged in the women's and peace movements, among others as cofounder and president of the *Neuer Wiener Frauenklub* (New Vienna Women's Club), founded in 1903.

From 1908 to 1909, Yella Hertzka received advanced horticultural training at the horticulture school for women in Bad Godesberg near Bonn in Germany, already considering establishing a similar institution in Vienna (C. Oesch, pers. commun 2009). Back in Vienna, Hertzka set up her horticulture school next to a villa colony that she and her husband had initiated. Up to 1938, Hertzka was the head of the horticultural school for women and taught floriculture, soil science, business operations, and law.

[5] Waschnitius, born in 1885 in Smichov (now part of Prague), enrolled as a guest student, which was the only possibility for women to attend a horticulture school at that time. The situation was similar to horticultural high schools in Germany, e.g., the *Königliche Gärtnerlehranstalt* in Berlin–Dahlem, which accepted women as guest students beginning in 1907 (Schekahn 2000: 74). After her graduation, Luise Waschnitius worked as an assistant at the *Pflanzenzüchtungsanstalt Mendeleum* (Plant Breeding Institute Mendeleum) in Eisgrub (today: Lednice, CZ). In 1917, she married Eduard Rossi, which is the last piece of evidence identified.

[6] In 1889, Lady Warwick founded the *Horticultural College for Women* in Reading, Berkshire. In Germany, Hedwig Heyl founded the first horticulture school for women in 1890 in Charlottenburg (at the time a city west of Berlin, now incorporated). Subsequently, others were opened in Friedenau (1894), Holtenau bei Kiel (1899), Bad Godesberg (1904), Wolfenbüttel (1905), Weimar (1908), Kaiserswerth am Rhein (1909), and Marienhöhe bei Plön/Holstein (around 1910) (Schekahn 2000: 76–86).

To register at Hertzka's school, students had to be at least 16 years old and have graduated from a secondary school. In contrast to Eisgrub, students were not obliged to have served a horticultural apprenticeship. Similar to horticulture schools for women in Germany, the school had two grades and aimed at providing a solid horticultural and business education. The school's final exam was considered—at the minimum—the equivalent of a full apprenticeship.

Yella Hertzka's school became the leading training institution for women in horticulture and garden architecture in Vienna and the former Austro-Hungarian Monarchy in the early twentieth century (M. A. 1920). In opening a horticulture school, Hertzka offered many women the potential to attain a higher professional training and a basis for economic independence as gardeners or garden architects.[7] Thus, most of the very few women (but hardly any of the men) who worked as garden architects in Vienna in the early twentieth century came from the liberal Jewish bourgeoisie (see Krippner and Meder 2011).

The *Villenkolonie Kaasgraben*, in the Vienna villa district of Grinzing, was initiated by Emil and Yella Hertzka and designed by Josef Hoffmann in the vocabulary of late Secessionist architecture. In the *Österreichischer Werkbund*, founded the same year as the *Villenkolonie*, 1912, following the *Deutscher Werkbund* founded in 1907, garden architecture also played an important role. The Hertzkas, who were friends of Gustav Mahler, Arnold Schönberg, Alban Berg, and Ernst Krenek, had originally envisioned the *Villenkolonie*, in the beginning called *Villenkolonie Hertzka*, as a colony for composers and people from the music business. In the end, composers like Egon Wellesz and intellectuals like the social–democratic politician Hans Vetter, cofounder of the *Österreichischer Werkbund*, moved in. The gardens of the villas, including a particularly large one at Hans Vetter's house, were designed by Josef Hoffmann himself in the formal, geometrical shapes of the *Modern Garden* style, which was developed by Secessionist architects and garden architects such as Franz Lebisch, Joseph Maria Olbrich, and Hoffmann in the first years of the century.

3 Housing Concepts of the *Wiener Schule*

Unlike for the Vienna Secession's aestheticist approach, for the *Wiener Schule*—with protagonists including Adolf Loos, Oskar Strnad, and Josef Frank—following an ideal of cultivated nonchalance and the direct relationship of house and garden in the tradition of the English country house was essential. A characteristic example is Strnad's Vienna house for the German writer Jakob Wassermann and his wife Julie (1914), showing the dynamic *Wiener Schule* concept of initiating a

[7] From 1912 to 1927, 180 women gardeners were trained at her school (Anonymous 1927), whereas in Eisgrub, during the same period, only 52 out of 334 graduates were women (Recht 1976).

continuing process of change and integration of the built construction into the vegetation at the moment the clients take up residence in it. The building is supposed to become an integral part of its surroundings. The small house, forming a *Wohnhof* (outdoor living space) as part of the living room by its L-shape, is characterized by an attitude of improvization, changeability, and even randomness, and seems to be permanently modified, therefore always imperfect, and gradually merged with its surroundings. This concept is also in evidence in the perennials and climbing plants that seem to have grown without prodding on the walls, but whose positions are precisely set forth in the garden plans (see Meder 2007).

A *Villenkolonie* like that of the Hertzkas was a common feature at that time. In 1913, Josef Frank drafted a small project of four flat-roofed terraced houses in the hilly outskirts of Vienna. In the end, only two houses with gardens on the back side of the small plots were realized. One of them was commissioned by Oskar Strauß, owner of a textile-trading agency, and his wife Hanny. Hanny Strauß ran a perennial nursery, named *Windmühlhöhe* after the area where they lived, holding a remarkable 400,000 plants by the early 1930s (Meder 2008; Karner 2011). As a garden architect, Hanny Strauß worked with Josef Frank several times (see below).

The architect Margarete Schütte-Lihotzky recalled—without elaborating upon them—the "embittered discussions that these houses sparked in Vienna" (Schütte-Lihotzky 1985). These houses, along with Oskar Strnad's Wassermann Residence and Adolf Loos's Scheu Residence, were the first villas in Vienna with flat roofs. Aside from this fact, the tentative, non-representational design of the houses may also have contributed to such discussions. Frank's goal was to erect an anonymous architecture of the everyday—buildings that tend to speak quietly and without specifics to their surroundings, yet were tailored to their residents were modifiable in essence and in shape. Personal "uncalculable contingencies"[8] (Frank 1919: 412) were deliberately afforded space. Frank conceived of the single-family house as living organism, which is, in response to the changing needs of their residents, ultimately subject to continual change (Frank 1919). The impression that the house has grown and been grown upon is characteristic of houses such as at Strnad's Wassermann Residence, where the vegetation on the street and garden elevations is to inhabit a climbing trellis: Frank drew the dwelling's sides facing the street and the garden covered with climbing plants and surrounded by sunflowers. On the ground level, direct access to the garden from the living room guarantees a close relationship of the houses to the outdoors.

4 The Garden of the Modern House

After World War I, in "Red Vienna," now governed by the social democrats, architects such as Frank, Loos, and Margarete Schütte-Lihotzky worked, for example, with the garden architects like Albert Esch when, inspired by Camillo Sitte and

---

8 "unberechenbaren Zufälligkeiten."

Leberecht Migge, they designed the settlers' gardens and garden courtyards of communal housing.[9] For the *Wiener Schule*, the close relation of inside and outside had special significance. In Josef Frank's workers' settlement in Ortmann-Pernitz, Lower Austria (1920–1923), whose gardens Esch designed, "in order to make the contact between house and garden as close as possible"[10] (Frank 1924), the combined kitchen and living room, occupying the entire depth of the house, opens directly on to the garden by means of a glazed French door.

In 1925, Frank and the architect Oskar Wlach established a home furnishing business named *Haus und Garten* that offered not only interior design and furniture, but also garden design. An example of the company's *Wiener Schule* garden architecture is Frank's Vienna house for the rubber manufacturers Julius and Margarete Beer (1930); he refers to it in his programmatic article *Das Haus als Weg und Platz* (The house as street and square). Frank describes his concept of a heterogeneous layout of living room and garden that can only be fully appreciated by moving about them.

The secluded character of the street facade dissolves in terraces and a looser spatial composition extending several levels on the garden side; the small corner balcony's slender columns are an allusion to the trunks of the trees. The outward multiplicity of the shapes of space in the context of inside and outside, and open and closed corresponds to the diversity of relations and dimensions of space inside the building. The crux of the design is the heightening—by means of spatial composition and employment of light—of the circulation. Frank's informally arranged, centrifugal space zones are meant to be experienced through movement in and through them. Since the paths are not axial, each room also has secluded zones. Strnad's and Frank's open systems do not claim conclusiveness but deal with their own contradictoriness. In ambivalent surroundings, the architects provide "open worlds" (Offene Welten) (Strnad 1922). The design of the house cannot be understood from a single standpoint, but in a succession of partial aspects relating to each other. Frank applied Camillo Sitte's urbanistic concept of street and square[11] to the layout of the house and its garden following Leon Battista Alberti's notion of *domus minima civitas* (the house as a small town).[12] The duality of the notions of street and square with its implication of rest and movement, statics and dynamics, is essential for Frank's work.

---

[9] Successful Vienna garden architects like Albert Esch or Willi Vietsch usually worked together both with modern architects like Josef Frank, Hofmann/Augenfeld, and Margarete Schütte-Lihotzky and with most conservative ones like Felix Angelo Pollak and Rudolf Perthen. In the early 1920s, e.g., Esch designed a weekend garden with Schütte-Lihotzky (see Schmidt 1995; Berger 2005).

[10] "[...] den Kontakt zwischen Haus und Garten möglichst innig zu gestalten."

[11] This principle of articulating the path in curved or broken lines, only tangentially touching rooms or areas regarded as closed spaces, is a direct correlation with the urbanistic principles of Camillo Sitte. In his book *City Planning According to Artistic Principles* (Der Städtebau nach seinen künstlerischen Grundsätzen), published in 1889, Sitte propagated asymmetrical, irregular shapes of squares and streets, following the examples of medieval cities. Sitte's writings were basic both for the residential settlements of *Red Vienna* and for the *Wiener Schule* (see Meder 2009).

[12] In 1910, Frank had written his dissertation about Alberti.

5 Cooperation Between Architects and Garden Architects

In 1927, Frank was commissioned to design a winter garden for the family house belonging to Robert and Anna Lang. This is how the garden architect Anna Lang first met Ernst Plischke, who, as Frank's assistant, was responsible for redesigning her house. It was the beginning of a fruitful teamwork. Anna Lang had attended the Vienna *Kunstgewerbeschule* and then received horticultural training at the famous Rothschild's gardens. One of the Anna Lang's first works was the garden of the Mühlbauer Residence in Vienna, designed by Ernst Plischke, in 1932 (profil 1934: 65–67/103–105; Ottillinger and Sarnitz 2003: 237). In the sense of the *Wohngarten* style, Anna Lang saw gardens as enlarged living spaces and, thus, arranged the outer rooms in close relationship to the building.

A differentiation of layers of space is also characteristic of Ernst Plischke's Gamerith Residence, which was conceived for the painter Walter Gamerith in Seewalchen on Attersee (Upper Austria). The single-story wooden house stands on a platform on a hill overlooking the lake. The floor deck forms a panoramic terrace, covered by a flat roof and with the rhythm of delicate white beams flush with the edge, creating an in-between zone which is both inside and outside. Seen from the living room, the roof frames the view over the lake. The elaborate mise-en-scène of the view was fixed on site so that the panorama as seen through the full-length ribbon window, Plischke reports, "is divided in about one third each of sky, lake and mountains"[13] (Plischke 1935: 582).

In 1932, Anna Lang was commissioned to redesign and plant the garden of Hans and Anny Moller (Sarnitz 2003: 67–69). Their house and garden had been designed by Adolf Loos four years earlier. Loos, however, was not particularly interested in garden architecture. In connection with the garden's redesign, the architects Franz Singer and Friedl Dicker created a pavilion and new garden paths. Dicker and Singer had been fellow students of Anny Moller at Johannes Itten's private art school in Vienna, followed Itten to the Weimar Bauhaus in 1919 and established an architectural practice after their return to Vienna in 1923. Anna Lang put exuberant flowerbeds alongside the paths and arranged small trees and stepping stone paths on the lawn in order to loosen the orthogonally structured plot.

The unity of house and garden stayed the most important task for garden architects. In further development of a functional, sober modernism, the focus was the spatial conception of the garden in analogy to the house. In 1932, the garden architect Paula Fürth defined her notion of a contemporary garden: "The modern garden corresponds to the modern house: functionality, comfort, simplicity and upkeeping are its ultimate aims"[14] (Fürth 1932: 29). Moreover, "The house and its surroundings, in which the garden is to be integrated, are the sources from which the garden designer derives the basic idea of his work. In the closest connection of these two, the garden

---

[13] "[...] ungefähr in je ein Drittel Himmel, See und Gebirge aufgeteilt wird."

[14] "Der moderne Garten entspricht dem modernen Hause: Zweckmäßigkeit, Wohnlichkeit, Einfachheit und Instandhaltung sind seine obersten Gesetze."

develops as a coherent entity where paths and patios, lawns, trees, shrubs and flowers do not stand for themselves but get their raison d'être only from the entire context and only achieve their full effect in it"[15] (Fürth 1937). Designing *Wohngärten* in the 1930s, architects and garden architects cooperated more and more—a development coinciding with the increasing professionalization of garden architecture. Thus, Fürth states that "today [...] through the cooperation of architect and garden architect, the house and garden form an organic union"[16] (Fürth 1937). The Fürth family had immigrated to Vienna from South Bohemia shortly after 1900 and was close friends with the family of Sigmund Freud. Paula Fürth studied natural sciences at Vienna University. In 1921, she earned a PhD degree for her dissertation "On the biology and microchemistry of some species of *Pirola*" (Zur Biologie und Mikrochemie einiger *Pirola*-Arten). She ran a nursery, a flower shop, a garden architecture studio, and the *Gartenbauschule Döbling* (Horticulture School Döbling) (Fürth 1930a, b) in the house that belonged to her family. The front of the flower shop was designed by the Jewish architect Fritz Rosenbaum, a close follower of Josef Frank.

In summer 1930, Fürth and Rosenbaum worked together for the Werkbund Exhibition at the *Österreichisches Museum für Kunst und Industrie* (Austrian Museum for Art and Industry). Fürth was responsible for the interior design and the flower equipment of Rosenbaum's model flower shop (see Born 1930a). In the same year, she designed a small indoor garden for an interior by the Jewish architect Liane Zimbler, at the exhibition *Wie sieht die Frau* (How does the woman see) in the Vienna Hofburg (Österr. Kunst 1930, no. 8: 11; 1932, no. 7: 29; archive of *Wiener Frauenkunst*, VBKÖ, Vienna). The exhibition was organized by the *Wiener Frauenkunst*, an association of Viennese women artists established in 1926 as a liberal secession of the *Vereinigung bildender Künstlerinnen Österreichs* (Association of Austrian Women Artists). Professional networking was an important strategy to promote women's careers in horticulture and garden architecture. They established their own professional associations, e.g., the *Verein der Grinzinger Gärtnerinnen*, and were also involved in existing ones oriented to art and architecture.

6 The Garden in the *Siedlung*

In 1932, Frank, at that time vice president of the *Österreichischer Werkbund*, was the artistic director of the Vienna *Werkbundsiedlung*. Hanny Strauß's nursery was

[15] "Das Haus und die Umgebung, in welche der Garten sich einzufügen hat, sind die Quellen, aus denen der Gartengestalter den leitenden Gedanken für sein Werk schöpft. In engstem Zusammenhang mit diesen beiden entsteht der Garten als einheitliches Gebilde, in welchem Wege und Sitzplätze, Rasenflächen, Bäume, Sträucher und Blumen nicht für sich selbst stehen, sondern nur aus dem Zusammenhang mit dem Ganzen ihre Daseinsberechtigung ziehen und nur in diesem zur vollen Geltung gelangen."

[16] "[dass] heute [...] Haus und Garten durch das gemeinsame Schaffen von Bau- und Gartenarchitekt als eine organische Einheit entstehen."

commissioned to deliver the plants for the gardens of the houses designed by Frank, Oskar Wlach, Hugo Gorge, and Oskar Strnad. Strnad, quoted by the architecture critic Else Hofmann, described the plantings as follows: "The beautiful, flourishing perennial garden Mrs. Hanny Strauß has designed here makes me very happy [...]. Through it an old dream of mine has been fulfilled: the flowers come straight into the apartment"[17] (Hofmann 1932: 26).

One of the row houses designed by the architect Richard Bauer in the *Werkbundsiedlung* was rented by the garden architect Willy Vietsch, who also redesigned three more gardens in the *Siedlung* (see Meder 2012). Vietsch and his colleague Wilhelm Hartwich had come to Vienna from Germany in 1925. They founded Hartwich and Vietsch, a nursery and garden design office; Vietsch was the designer. In the beginning, they took on the planting and took care of gardens Albert Esch had designed, e.g., the garden of the *Hietzinger Schule*, a private school with psychoanalytically trained teachers founded by Eva and Valentin Rosenfeld in the backyard of their Vienna house, which had been redesigned by Adolf Loos (see Estate of Valentin and Eva Rosenfeld).[18] Soon, however, as a successful garden architect, Vietsch worked with architects like Karl Hofmann and Felix Augenfeld (Vietsch 1935a, b, c, d, 1936), former students of Adolf Loos and assistants of Oskar Strnad.

Vietsch wrote numerous articles on garden design, covering topics ranging from planting to advertising and professional standards. In them, he presented himself as proponent of the settlers' movement surrounding Otto Neurath, Adolf Loos, Josef Frank, and Margarete Schütte-Lihotzky, who rejected multi-story apartment buildings, such as those predominantly built by *Red Vienna* (see e.g., Vietsch 1931), and spoke out for a garden as outdoor living space to add to the residents' comfort: "In a time of sports and hygiene, the garden takes on increased significance. It is no longer to the same degree about representation, but now above all also should function as part of the dwelling"[19] (Hartwich and Vietsch 1930: n. p.). The aim was house and garden as coherent entity: "House and garden should not only be

[17] "Der schöne, blühende Staudengarten, den Frau Hanny Strauß hier angelegt hat, macht mich ganz glücklich [...]. Es ist mir dadurch ein alter Traum erfüllt worden. Die Blumen kommen ge—radewegs in die Wohnung."

[18] Adolf Loos also cooperated with Albert Esch in 1923 when Esch designed the garden of the Spanner Country House in Gumpoldskirchen, south of Vienna. When Loos and his office partner Heinrich Kulka designed the Khuner Country House in Payerbach in the Semmering region in Lower Austria in 1930 (Kulka 1979: 43), the spaces surrounding it were designed by Grete Salzer, the client's sister-in-law (Austrian State Archive, Vienna, Aryanization Files, 1938). Beside Yella Hertzka, whose school Grete Salzer had attended, and Paula Fürth, Salzer was the third Jewish woman to run a private horticultural school in Vienna, only a few houses away from Paula Fürth's school, plus an office for garden design and a nursery named *Hortensium*, on the plot of her parents' house, similar to the likewise unmarried Paula Fürth.

[19] "In der Zeit des Sports und der Hygiene kommt auch dem Garten eine erhöhte Bedeutung zu. Er dient nicht mehr in gleichem Maß wie früher der Repräsentation, sondern vor allem auch den Zwecken des Wohnens."

fully unified spatially and formally, but also intellectually—because it takes a beautiful garden to give our home its outer frame, its well-structured framework. [...] If the garden is to satisfy our contemporary notions, we will have to bring it into close contact with our living spaces"[20] (Hartwich and Vietsch 1927: 6).

In the *Werkbundsiedlung*, Vietsch designed the gardens of the two row houses designed by Margarete Schütte-Lihotzky (see Vietsch 1933, 1935b). In his text on his own garden—published following Austria's "Anschluss" to Nazi Germany—which defended the *Werkbundsiedlung* and its architecture, Vietsch built on the Viennese modernism of Josef Frank and his like-minded colleagues and their notion of a connection between house and garden: "The spacious patio adjoining the living room in my little garden is practical and a nice view. Such a patio is in the truest sense of the word an extension of the dwelling. [...] The direct link between living room and garden has a pleasant effect. That's the beauty of it: if one can enter the garden without first using a stair or a side entrance. This possibility should be taken into consideration more often by master builders and clients"[21] (Vietsch 1938, see also 1936).

Following internal conflicts, the *Österreichischer Werkbund* split soon after the *Werkbundsiedlung.* In 1934, the *Neuer Werkbund Österreichs* (New Werkbund of Austria) was founded against the (Jewish) "internationalists" around Josef Frank by conservative members like Josef Hoffmann, Peter Behrens, and Clemens Holzmeister (see Posch 2010). It is remarkable that Hanny Strauß designed an indoor garden at the *Neuer Werkbund Österreichs* Christmas Exhibition in 1935, an organically shaped elevated flowerbed serving as a partition (Simony 1936). At the Paris World Fair of 1937, Strauß created the outer space and the courtyard garden of the Austrian pavilion that Oswald Haerdtl had designed (see Selinko 1937: 2–3). In direct connection to the garden courtyard with free, floating forms and a water basin, Josef Frank's *Haus und Garten* designed a modern living room.

Just a year later, after the "Anschluss," the *Werkbund* was liquidated. Hanny Strauß, Paula Fürth, Yella Hertzka, and Grete Salzer had to flee Austria. Many architects like Josef Frank, Franz Singer, Oskar Wlach, Heinrich Kulka, Liane Zimbler, Karl Hofmann, and Felix Augenfeld left the country too, as did most of their clients, like the Moller, Khuner, Rosenfeld, and Beer families. Non-Jewish

[20] "Haus und Garten sollen nicht nur räumlich und formal, sondern auch geistig ein einheitliches Ganzes bilden—denn erst der schöne Garten gibt unserem Heim den äußeren Rahmen, seine formvolle Fassung. [...] Soll nun der Garten unseren Zeitansprüchen gerecht werden, so müssen wir denselben eng in Zusammenhang und Verbindung mit unseren Wohnräumen bringen."

[21] "Praktisch und auch in der Ansicht nett ist in meinem Gärtchen die geräumige Sitzterrasse vor dem Wohnzimmer. Eine solche Sitzterrasse ist im eigentlichen Sinn eine Erweiterung der Wohnung. [...] Angenehm wirkt die unmittelbare Verbindung des Wohnzimmers mit dem Garten. Das ist ja gerade das Schöne, wenn man, ohne erst über Stiegen oder Nebenausgänge zu gehen, in den Garten gelangen kann. Diese Möglichkeit sollte viel mehr als bisher von den Baumeistern und von den Bauherren berücksichtigt werden." Another garden, this one in connection with a house designed by Oskar Wlach, was laid out in accordance with Vietsch's design by the residents themselves (see Gälzer and Posch 1994; Meder 2012).

Margarete Schütte-Lihotzky went back to Austria from her Turkish exile in order to join the resistance movement and was subsequently sentenced by a Nazi court to life imprisonment. Friedl Dicker refused to leave without her husband and was killed in a concentration camp. Non-Jewish Willi Vietsch, a socialist like Schütte-Lihotzky, took his life in 1944 when he was ordered to join the *Wehrmacht*. The violent expulsion of Jewish and socialist architects and garden architects from Austria was a tremendous loss for the country's architecture and garden design, and its culture as a whole. Not least of these was the end of the programmatic pursuit of the unity of house and garden; after World War II, it has not been possible to resume it.

References

Anonymous (1906) Papst Pius X. und die Gartenbauschulen für Frauen in Österreich. *Gärtnerische Rundschau*, no. 5:1
Anonymous (1927) 15 Jahre Gartenbauschule für Frauen, Wien XIX, Kaasgraben. Gartenzeitung no. 8: 142
Augenfeld F (1981) Erinnerungen an Adolf Loos. *Bauwelt,* 1907
Beller S (1989) *Vienna and the Jews: 1867–1938; a cultural history*. Cambridge University Press, Cambridge
Berger E (2005) "… Gärten zu schaffen, die Schönheit und Nutzen vereinen …" Albert Esch (1883–1954). Die Gartenkunst 1:22–72
Boeckl M (ed) (1995) Visionäre and Vertriebene. Österreichische Spuren in der amerikanischen Architektur. Exhibition catalogue Kunsthalle Wien
Born W (1930a) Ausstellung des österreichischen Werkbundes in Wien. Deutsche Kunst und Dekoration 66:305–310
Born W (1930b) Schöne Wiener Privatgärten. Die Bühne 285:29–30
Eisler M (1936) Oskar Strnad. Gerlach + Wiedling, Vienna
Enis R (2006) Zionist Pioneer women and their contribution to garden culture in Palestine, 1908–1948. In: Inhetveen H, Schmitt M (eds) Frauen und Hortikultur. LIT Verlag, Hamburg, pp 87–114
Fliedl G (1986) Kunst und Lehre am Beginn der Moderne. Die Wiener Kunstgewerbeschule 1867–1918. Residenz, Salzburg, Vienna
Frank J (1919) Das neuzeitliche Landhaus. Innendekoration, 412
Frank J (1924) Die Wiener Siedlung. Der Neubau, 28
Frank J (1931) Das Haus als Weg und Platz. Der Baumeister, 316–323
Fürth P (1930a) Wiener Gartenbauschulen. Die Bühne 284:26–28
Fürth P (1930b) Der moderne Garten. Architektur und Bautechnik, 78–79
Fürth P (1932) Gärtnerinnen sprechen über ihre Gärten. Österr Kunst 7:29–30
Fürth P (1937) Stein und Beton als Gartenwerkstoffe. Innendekoration 1937:177
Gälzer R, Posch W (1994) Die Gärten der Wiener Werkbundsiedlung 1932. In: Schmidt E, Hansmann W, Gamer J (eds) Garten—Kunst—Geschichte. Festschrift für Dieter Hennebo. Worms 180–186
Hartwich und Vietsch (1927) Unsere Stauden im Garten. Authors' publishing, Vienna
Hartwich und Vietsch (1930) Wie gestalte ich meinen Garten? Authors' publishing, Vienna
Hecht L (2007) Jüdische Frauen zwischen Emanzipation und Tradition. In: Kohlbauer-Fritz G, Krohn W (eds) Beste aller Frauen. Weibliche Dimensionen im Judentum. Exhibition catalogue Jüdisches Museum Wien/Holzhausen, pp 144–156
Hofmann E (1932) Hanny Strauß—die Staudengärtnerin. Österr Kunst 7:26–27
Karner E (2011) Hanny Strauss und die Stauden—eine Liebesgeschichte. Historische Gärten 1: 4–9

Karner E (2012) Zwischen Gartenbau und Gartenkunst: Gärtner und Gartengestalter in Wien 1918–1945. Die Standesgeschichte im Wechsel der politischen Systeme. PhD thesis, Vienna, Technische Universität

Krippner U, Lička L (2008) "A Garden for Pleasure." Die Gartenarchitektin Anna Plischke (1895–1983) und ihre Werke in Wien und Wellington. In: Fischer H, Wolschke-Bulmahn J (eds) Gärten und Parks im Leben der jüdischen Bevölkerung nach 1933. Meidenbauer, Munich, pp 365–384

Krippner U, Meder I (2011) Cultivating, designing, and teaching. Jewish women in modern Viennese garden architecture. Landscape Res 36(6):657–668

Kulka H (1979) Adolf Loos. Löcker, Vienna (1930)

M A (Aull M) (1920) Bericht über die Tätigkeit der Gartenbauschule für Frauen in Grinzing seit Beginn des Weltkrieges. Z. für Garten- u. Obstbau. 1. Gartenbau, Gartenkunst 34–35

Meder I (2003) Vertrieben, Ermordet, Vergessen. Architekten der Wiener Schule und ihre Bauherren in Tschechien. Architektur Aktuell 4

Meder I (2004) Offene Welten—die Wiener Schule im Einfamilienhausbau 1910–1938. PhD thesis, Stuttgart University

Meder I (ed) (2007) Oskar Strnad 1879–1935. Exhibition catalogue Jüdisches Museum Wien. Pustet, Salzburg

Meder I (ed) (2008) Josef Frank—eine Moderne der Unordnung. Pustet, Salzburg

Meder I (2009) House and exterior in the architecture of the "Vienna School". In: Landscape—Great Idea! X-LArch III. 29.April-1.Mai 2009, Univ.f. Bodenkultur, Vienna

Meder I (2010) Lebens- und Arbeitsbedingungen jüdischer Architekten in Österreich. In: Senarclens de Grancy A, Zettelbauer H (eds) Architektur. Vergessen. Jüdische Architekten in Graz. Böhlau, Vienna, pp 69–75

Meder I (2012) "Natur und Architektur werden hier ineinandergeschoben"—Haus und Garten in der Werkbundsiedlung. In: Ein Manifest des Neuen Wohnens. Werkbundsiedlung Wien 1932. Exhibition Catalogue, Wien Museum, pp 96–101

Meder I, Krippner U (2011) "Die Blumen kommen geradewegs in die Wohnung"—Gartenarchitektur im Österreichischen Werkbund. In: Butenschön S (ed) Garten—Kultur—Geschichte. Gartenhistorisches Forschungskolloquium 2010. Technical University Berlin, pp 55–59

Meredith A (2003) Horticultural education in England, 1900–1940: middle-class women and private gardening schools. Garden History 31(1):67–79

Oesch C (2008) Yella Hertzka (1873–1948). Eine Auto/Biographie von Beziehungen. In: Gehmacher J, Hauch G (eds) Auto/Biographie, Gewalt und Geschlecht. Österreich. Z. f. Geschichtswissen. Studien Verlag, Innsbruck, pp 118–144

Österr. Kunst (1930), no. 8: 11; 1932, no. 7: 29

Ottillinger E, Sarnitz A (2003) Ernst Plischke. Prestel, Munich

Pass Freidenreich H (2002) Female, Jewish, and Educated. The lives of central European university women. Indiana University Press, Bloomington

Plischke E (1935) Ein Haus am Attersee. profil 582–584

Posch W (2010) Clemens Holzmeister. Müry Salzmann, Salzburg

Profil (1934) 65–67, 103–105

Raggam-Blesch M (2008) Zwischen Ost und West. Identitätskonstruktionen jüdischer Frauen in Wien. Studien Verlag, Innsbruck

Recht H (1976) Die höhere Obst- und Gartenbauschule und das Mendeleum in Eisgrub. Verlag des wissenschaftlichen Antiquariats H. Geyer, Vienna

Sapák J (2000) Brno's Jewish Architects. In: Brněnští židovští architekti—Brno's Jewish Architects 1919–1939, 8–19. Obecní dům, Brno

Sarnitz A (2003) Adolf Loos 1870–1933. Taschen, Cologne

Schekahn A (2000) Spurensuche 1700–1933. Frauen in der Disziplingeschichte der Freiraum- und Landschaftsplanung. Universität Gesamthochschule Kassel

Schmidt S (1995) Albert Esch: Ein österreichischer Gartenarchitekt der ersten Republik. Die Gartenkunst 7:309–317

Schütte-Lihotzky M (1985) Erinnerungen an Josef Frank. Bauwelt 1052
Selinko A (1937) Interview mit Oswald Haerdtl. Die Bühne 452:2–3
Simony S (1936) Der neue Werkbund Österreichs stellt aus. profil 1:24–28
Strnad O (1922) Neue Wege in der Wohnraumeinrichtung. Innendekoration 323
Theimer C (1909) Frauenarbeit in Österreich. Opitz, Vienna
Vietsch W (1931) Über Gartenlösung (sic!) der Siedlung 'Tivoli' und 'Fasangartenstraße'. Eigenheim und Weekend, Jan 12
Vietsch W (1933) Ein Garten in der Werkbundsiedlung. Mein Garten 10:157–158
Vietsch W (1935a) Gartensitzplätze. Österr Kunst 9:24–25
Vietsch W (1935b) Ein Reihenhausgarten. Die Gartenschönheit 273
Vietsch W (1935c) Umgestaltung zum Wohngarten. Die Gartenschönheit, pp 26–28
Vietsch W (1935/36) Der farbenfrohe Garten. Die Pause 11:36
Vietsch W (1936) Kleiner Wohn- und Nutzgarten. Nach der Arbeit 48:7–9
Vietsch W (1937) Der Weg zum schönen Garten. Das Wüstenroter Eigenheim 250
Vietsch W (1938) Mein Garten. Nach der Arbeit 27:568–569
Wiener Architekten. 1935, vol 14 Arnold Karplus, Gerhard Karplus. Elbemühl, Vienna
Wischenbart R (1992) "Nirgends Türen; alles fest verschlossen". Die Zeit 25(9):48

Archive Material

Archive of "Wiener Frauenkunst", VBKÖ, Vienna
Austrian State Archive, Vienna, Aryanization Files, 1938
Estate of Valentin and Eva Rosenfeld, Sigmund Freud Museum, Vienna

Green Walls

Maria Bostenaru Dan

Abstract In this paper, several steps for our research, in which the green walls are dealt with (documentation of the spread of green walls in Europe and housing by us with interior and exterior green walls), as well as continuous approaches (such as an invention patent for supporting green wall system), and in general the land art (contribution to the reconstruction of L'Aquila or simply representing the concept of space) are presented. Employing our experience, we will help the virtual and real land art work management, as well as contributing with two creations of us: virtual houses with green walls, which can be presented in real-time applications, and available on the Internet, as well as an alternative photographic exhibition in which the green plant materials as well as Art Nouveau wall drawings come to a dialogue. The research will help to highlight the contribution of green walls to landscape design and to the dissemination of this knowledge, which is currently not enjoying so much awareness. Increased awareness will help employment in post-disaster situations, which is the author's area of research.

Keywords Green wall • Pocket park • Geotextile • Interwar • Public participation • Art nouveau • Cost • Earthquake • Multimedia • Installation • Interior

1 Introduction

Bucharest, the capital of Romania, is a city lacking green spaces. In the interwar time, there have been attempts to create a green belt, but they were given up. Gardens in the middle of the city are getting lost in time. At the same time, there

M. Bostenaru Dan (✉)
Urban and Landscape Design Department and Centre for Architectural and Urban Studies, "Ion Mincu" University of Architecture and Urbanism, Academiei 18-20, 010014 Bucharest, Romania
e-mail: maria.bostenaru@iaim.ro

C. Crăciun and M. Bostenaru Dan (eds.), *Planning and Designing Sustainable and Resilient Landscapes*, Springer Geography, DOI: 10.1007/978-94-017-8536-5_11,

are numerous parcels empty with large brand walls adjacent to them. There is no research on how to make these usable at this particular site, although "pocket parks", the issue of making parks on empty parcels, is an international topic with dedicated publications since the 1990s (e.g. Pocket Parks in Barcelona or the US).

Object of the investigation is using rest surfaces in the city. The city is separated from the countryside, a process which happens since centuries. The nature remains outside. Which is the role of green spaces in shrinking or growing cities? From here arises the first research question: where are we in solving the problem of separation of the urban and rural area, which are the potentials to solve this problem?

There is vast literature regarding green roofs and green walls, databases like SCOPUS listing more than 1,000 titles. In many cases, geotextiles are used, but the studies relate mainly to energy saving, not to emergency. We will do a review of the most significant ones during our research. The research questions connected to this are about the history of green walls, about the concepts and their employment till now. Which has been the perception of green walls so far? Which is the relationship between employment of green walls and the habits of living in everyday life? A question arising from literature is what do green walls cost. This is the financial benefit of green spaces in raising the value of the parcels adjacent to them compared to construction on those parcels.

We did a review of implemented examples of green walls. What we aim with this is to find examples of best practice. The research question here relates to the process of transformation towards healthy spaces.

2 Research Methods

As research methods, a combination of field research, literature research and research by design has been employed.

How regression, a method of Joedicke (1976), is applied to decision processes, including participation was presented in Bostenaru Dan (2004): "It is an approach where characteristics and criteria are deducted based on analysis of existing projects, while know-how is derived by means of scientific disciplines. In the first step of the deductive phase a general hypothesis had to be drawn out from the one formulated after induction. In a second step, statements for a particular case can be derived".

2.1 Site Visits

As already mentioned, a research method was the site visit and first-hand experience on the location of the object of investigation. Analysis was performed while recording in photography and surveying by drawing.

Fig. 1 Stones in Vietnam photo. Photocase/Daniel Schoenen

Fig. 2 Natural green walls and humidty: Sintra. *Photo* M. Bostenaru 2013

As Patrick Blanc presented at the Anuală exhibition in Bucharest 2011, green plants are growing on stone in Vietnam without earth, just by help of water and most probably nutrients (Fig. 1). The humid environment leads to the same in Sintra (Fig. 2), Portugal and in Stresa (Fig. 3), Italy, in the vicinity of the ocean and respectively a large lake.

Fig. 3 Natural green walls and humidty: Stresa. *Photo* M. Bostenaru 2012

We examined however the designed green walls, outgoing from those typical for Montreal where we designed an installation, going over the Patrick Blanc type walls, and creative combinations with pots or growing trees and canopy.

2.1.1 Quarters in Montreal

During our stay as support grantee at the Canadian Centre for Architecture in Montreal in 2010, there was a project called "green CCA" which however had little to do with green spaces and vegetation, and rather with ecology. However, we did an excursion to discover housing in the neighbourhoods of Montreal, which led to this kind of discoveries. Also, we did a private cultural tour of the buildings of Dan Hanganu.

Dan Hanganu is one of the architects whose works marked Montreal. At the archive centre he designed in 1999, there is a green wall with climbing plants, an alternative to his multimedia facades (Fig. 4). The dialogue between green and multimedia is a topic to which we will return later.

Figure 5 shows an example where the green climbing plants contribute to the expression of the tectonics of the building. Although the lower floors used to be with glass to express the transparence of public spaces since the Loos house in Vienna, this is opposite to Semper's (1863) tectonics of the foundation, wall and roof hierarchy. The weight of the massive upper floors is counterparted by the green growing on the glass, which offers massiveness also to these. The comparison to

Fig. 4 Montreal archive centre, project 1999, photo 2010, rue Labelle, centre. *Photo* M. Bostenaru

Fig. 5 Expression of function and expression of tectonics in Montreal neighbourhood. *Photo* M. Bostenaru 2010

Fig. 6 Growing trees in Montreal. *Photo* M. Bostenaru 2010

Gottfried Semper is essential also from another point of view. Semper developed the theory of clothing, according to which, walls (Wand in German) are comparable to the textile of the cloth (Gewand), and as such, the green wall of felt-based textile would express the clothing of a building best. Although we could discuss the textile paradigm outgoing from the one of Deleuze (1980) where the metaphor between felt and weaving is compared to that between flat and striated.

Figure 6 shows the natural growth of trees on walls in Montreal, similar to what Hundertwasser designed in Vienna, which we will see later in this paper.

Climbing plants are also to be found on the usual condo(minium) housing which are typical for low-cost residences in Montreal (Fig. 7).

2.1.2 Felt-Based Green Walls (Living Walls)

When designing green walls of plants growing on felt material, Patrick Blanc was inspired by the growing vegetation in tropical countries such as Vietnam. In such countries, plants grow on stone, without earth, and this was an important argument. All they need to grow is not the earth, but the vitamins in the earth, and water, and this can be done at felt materials with water including vitamins.

A typical example of the tropical environment is the wall from Madagascar built in the Aquarium in Genova (Fig. 8), in the rehabilitation of the port by the architect Renzo Piano. The location of the green wall in an aquarium also makes use of the metaphor of the water needed by the plants to grow. This is an interpretation at building scale of what for a city might mean the blue-green infrastructure, a popular concept nowadays.

Fig. 7 Typical condo(minium) housing in Montreal. *Photo* M. Bostenaru 2010

Another landmark by Patrick Blanc is the wall of the Caixa Forum in Madrid (Fig. 9). The green wall transforms into a luxury landscape, a wall which would have been otherwise blind, lacking windows and needing to be covered by another building. It is part of the rehabilitation of an industrial site. However, the problem of watering the plants has been solved unsatisfactorily.

Another approach to such an empty wall to which it is designed to attach a building has been done at Stilwerk in Vienna (Fig. 10). Instead of attaching the building to the adjacent one, Jean Novel built at a certain distance from it, and with a wall out of glass, allowing a view to the green wall designed by Patrick Blanc. It is a luxury view of some shops. We will see in one of the design projects we propose in the research by design section that we also approached the method of building a green wall at a certain distance from a building with windows, even in an empty space. It is a design method we followed in another competition projects as well

In Vienna, such green walls based on felt have entered vernacular architecture, as we can see at the entrance of a bar (Fig. 11).

The climate in Vienna is favourable for green walls, but even more is the climate in Lisbon, where we find a Mediteranean climate and the vicinity of the ocean. Lisbon is also neighbour to Sintra, where green walls grow naturally, due to the humid atmosphere, like on the shores of Lago Maggiore in Italy.

Large scale felt-based green walls we can find in Lisbon at the Natura Towers (Fig. 12), a project by Vertical Garden Design. Here, the green walls are both found in the height, separating tower models, as well as used for the plaza of the towers. The walls in the whole height show however lacking maintenance possibilities. The towers also feature an interior garden with a waterfall.

In Lisbon Patrick Blanc designed the green wall in the shopping mall "Dolce Vita" in the periphery of the city (Fig. 13).

Fig. 8 Aquario di Genova. Patrick Blank. Madagascar wall. *Photo* M. Bostenaru 2012

A small wall decorating the otherwise empty wall of the Café Royale in the centre is an example of how such an intervention can enhance the quality of spaces (Fig. 14).

Recently, a lot of publicity has been done for the house in Traversa do Patrocinio (Fig. 15), in the centre of Lisbon, close to the Parliament and the church of Estrela. Here, the architect followed similar principles of the façade as Patrick Blanc did for Musee d'Orsay in Paris, his best known work for green walls (see another contribution in this book). The green wall does not cover anymore a blind wall, but band windows are cut into it.

An original approach is the one in Basel, where instead of a wall, at the Museum of culture there are green plant hanging from the roof along the façade (Fig. 16). Patrick Blanc has an approach for hanging walls when he approaches

Fig. 9 Caixa forum. Madrid. *Photo* M. Bostenaru 2012

ceilings, for example the Electra exhibition in Paris (2007–2008) (http://www.verticalgardenpatrickblanc.com/realisations/paris/electra-edf-folies-vegetales) or the "serre du Museum national d'histoire naturelle", Paris (2010) (greenhouse of the natural history museum http://www.verticalgardenpatrickblanc.com/realisations/paris/serre-du-museum-national-dhistoire-naturelle-paris-vegetal-ceiling), vegetal ceiling. We will come back to this concept in our designs.

Fig. 10 Stilwerk tower. Vienna. Jean Nouvell. *Photo* M. Bostenaru 2012

Although spectacular, such green walls need proper watering systems, which might prove expensive for common buildings. If they are not designed properly, the plants might dry.

Fig. 11 Biergarten. Vienna. *Photo* M. Bostenaru 2012

2.1.3 Other Green Walls

Coming back to Vienna, the garbage administration at nr. 48 has been brought out of anonymity by a green wall designed by the University of Natural Resources and Life Sciences (BOKU) (Fig. 17). Instead of the felt green wall we talked about, here, plants are growing traditionally in their pots, just that there are rows of pots one over the other.

Another original approach marking Vienna and other cities is the green sustainable architecture promoted by Friedensreich Hundertwasser. Both the Hundertwasser

Fig. 12 Natura towers. Lisbon. *Photo* M. Bostenaru 2012

house and the Kunsthouse Vienna feature trees growing through the windows of the facades, and, as such green walls, not just the green roofs typical for his buildings (Fig. 18). To see how such an artistic show off project can come to dialogue with living habits, see the work of Kraftl (2009).

Climbing plants can be found also in Vienna, combined with the architecture of the Modern Movement, which might seem unusual, because the Modern Movement wanted to promote industrial new materials, and not the proximity to nature. Such one is a building by architect Josef Hoffmann (Fig. 19a). As an

Fig. 12 (continued)

opposite, in Lisbon, the architecture of the interwar times employed the approach by Hundertwasser, by creating interior courtyards at upper storeys on which trees can grow. Such one is the cinema theatre Eden by architect Cassiano Branco (1931–1932) (Fig. 19b) in Lisbon. The project of the cinema underwent more alternatives, the second (1930) not built, displaying a compact façade.

A similar approach to the one Hoffmann had in Modern Architecture we can see in Romania, in a house designed by the known Italian architect Gio Ponti in

Fig. 13 Dolce vita shopping mal. Lisbon. *Photo* M. Bostenaru 2012

Cluj Napoca (Fig. 20), a city where also Hoffmann designed some houses. The exterior of the house is fully covered by climbing plants.

Such climbing plants can grow over many floors, as in the foyer of the O_2 hotel in Dublin, having there the appearance of a felt-based green wall (Fig. 21).

2.2 *Literature Review*

A next step was literature research on the approaches used and archive research on the buildings in question.

Fig. 14 Café Royale. Lisbon. *Photo* of and by M. Bostenaru 2012

Research on green roofs is much more extended than on green walls. From these, we selected some paper which may give hints for green wall research.

Such one is Dvorak and Volder (2010) looking at the typology of green roof vegetation in North America, in a literature review. The paper was also useful for

Fig. 15 House on Traversa do Patrocinio. Lisbon. *Photo* M. Bostenaru 2012

Fig. 16 Green walls in Basel, Museum of Cultures. *Photo* M. Bostenaru 2013

us in providing criteria and a methodology for literature review. Another one is Scherba et al. (2011) modelling photovoltaic panels and green roof systems as alternative sustainable roof technologies. It provides insights into the difference between the active and passive energy savings design. Molineux et al. (2009) looked to growing media for green roofs out of recycle materials. Lundholm and Peck (2008) edited a special issue on green roofs as ecological systems from the Green Roofs for Healthy Cities annual conference. It features also the paper by Köhler (2008) on green walls, which we will later discuss. "Beth Ann Currie and Brad Bass adapt a widely used urban ecosystem model to generate predictions on the ability of various types of green roofs and walls to ameliorate air quality (Currie and Bass 2008). This paper was recently selected by the European Commission (an EU body) for inclusion as a Science and Policy news alert". Simmons et al. (2008) compare the different hydrologic and thermal performance of reflective and non-reflective green roofs. In the following, the differences between vertical greening systems will provide similar insights.

A key paper is Hindle (2012) which investigates the first designed of green walls, long before the concept spread through Patrick Blanc's patent. "Stanley Hart White, Professor of Landscape Architecture at the University of Illinois Urbana-Champaign from 1922 to 1959, was granted US Patent 2,113,523 on 5 April 1938 for the Vegetation-Bearing Architectonic Structure and System in which he describes the method for creating an 'architectonic structure of any buildable size, shape or height, whose visible or exposed surfaces may present

Fig. 17 BOKU design at 48, Vienna. *Photo* M. Bostenaru 2012

Fig. 18 Hundertwasser design at Kunsthaus Wien and at Hundertwasserhaus, Vienna. *Photo* M. Bostenaru 2012

Fig. 19 **a** Modern Movement house by architect Josef Hoffman house in Vienna (*side view* of Villa Knips 1924–1925) with climbing plants. **b** Interwar building by Modern Movement architect Casiano Branco in Lisbon with upper floor trees. *Photos* M. Bostenaru 2012

a permanently growing covering of vegetation'". (Hindle 2012). The paper contains drawings of the first concept for the Century of Progress Exhibition in 1933 as well as the patent drawings. "Patent application for the Vegetation-Bearing Cellular Structure and System was filed by William M. MacPherson on 4 April 1938" (Hindle 2012). MacPherson's patent presented the difference of having oblique cells. "Shortly after working with White and MacPherson on patent 2,113,523 and 2,121,173 attorney Elmer Hovenden Gates submitted his own patent for the Vegetation-Bearing Display Surface and System on 27 May 1938, and was granted US patent 2,279,735 on 14 April 1942".

Fig. 20 Casa Tataru. Cluj. Architect Gio Ponti. *Photo* M. Bostenaru 2012

An influential paper is Francis and Lorimer (2011). Living walls are identified in this paper as being an instrument for reconciliation ecology. Another key aspect identified in the paper is the role of public participation in promoting green walls, which is, as we will see, a different approach from the mostly technical ones of research on green walls, but one key one on green spaces. The paper also defines green facades and living walls. The main difference according to the article lays in the fact that in case of green facades, the plant grows at the base of the wall, while living walls are designed like vertical green roofs, with the support parallel to the wall. Different from the general research done on the energy aspects of vertical greenery systems, Francis and Lorimer (2011) mentions the indoor living walls, called biowalls, which are useful for psychological improvements of the indoor space, and to which we will return in our designs. Another key feature of living walls is that the range of allowed plants is larger, being able to be extended from the ones with the characteristics of climbing, and this may be a factor of urban reconciliation by allowing for rare species. However, there are biodiversity benefits and limitations which are investigated more in detail. We will return to this issue when dealing with the difference of drawing plants in Art Nouveau or via multimedia. It is also an issue of ecological engineering. A next issue investigated is the availability of space for living roofs and walls. It does not however identify specific parts of the building which are suitable (for example blind walls without windows) or the way composition might be created. Instead, it identifies socioeconomic barriers, which leads us to the next investigated article, by White and Gatersleben (2011) when talking about cultural perceptions. Economic aspects of green spaces in urban areas have been also investigated by us (Bostenaru Dan and Mendes 2013). Francis and Lorimer (2011) develop an adaptive citizen science model for living walls between researchers, professionals, government authorities and citizens as actors. The fact that creating ideas in the heads of citizens protects green spaces

Fig. 21 Dublin hotel with green wall. *Photo* M. Bostenaru 2012

better than any legislation is not new, but dating since the GreenBelt Frankfurt (Koenigs 1991).

Through its approach, claiming living (roofs and) walls to be ecological Francis and Lorimer (2011) called for a discussion in Henry and Frascaria-Lacoste (2012). In the discussion, it is agreed that private gardens can form patches as an interconnected network of green infrastructure. We show such an example in Chap. 12 of this work, authored by us. As the original paper, they also support the findings of Colding (2007). Green roofs as parts of the network are however questioned, and the discussion barely touches green walls. Unlike in this discussion, we support that these are part of the network and did the following reviews accordingly.

Generally, articles/papers on contemporary research regarding green walls concern the technical improvements brought by these. White and Gatersleben (2011) wrote one of the few articles on the psychological improvements. Through sociological research methods such as survey (online) and interviews, their research proves the preference for some types of houses with vegetation. Vegetation has been superimposed to photographs for the scope of this study and, in case of the ivy façade, presents the characteristics of climbing plants on walls with windows. The scope of the study was to see whether this kind of intervention shall be a viable option for restoration. Also Wang and Wu (2011) look at how green walls can contribute to the restoration of buildings.

Imam (2006) looks at an issue which might be derived from this: the role of greenway systems, for example linked parks, in planning residential communities. Case study is Cairo, Egypt. If we were to consider pocket parks part of this kind of linked systems, then green walls may be part of them.

Tachibana et al. (2011) identify exactly green walls as a means for the greening of urban areas. However, as we earlier noticed in our field research, plants may dry, leading to a landscape called by the authors "bad". Maintenance is therefore important. A table with examples is provided.

Xuesong and Lei (2011) looked in detail to this psychological effect, namely in the form of several individual large buildings in case of renovation: the green wall compared to media facades, and different combinations of green on the façade, which can lead to the design we later discuss.

Colding (2007) introduces the concept of "Ecological land-use complementation". According to landscape complementation, "in a landscape of different patch types, such as in heterogeneous, urban landscapes, a species needs to move between patches to obtain critical amounts of resources. […] In addition, ecological land-use complementation draws on the island biogeography theory" according to which "area is a key determinant for species occurrence and diversity and it is generally assumed that habitat diversity increases with area". Examples for this are urban green patches and agriculture. A guiding principle identified is "Identify and make use of existing ELC-structures in the landscape for the development of "arenas" for experimental design and adaptive co-management. The goal of such sites could be the working out of adaptive management policies that will best meet the various needs of critical species (e.g. focal species), as well as promoting Local Agenda 21 in city-regions". The idea of patches from the discussion on Francis and Lorimer (2011) returns.

Carter and Fowler (2008) looked at environmental policy instruments for establishing a green roof infrastructure. Green roofs are considered here only an example for "innovative practices which create ecosystem services and ecologically functional land cover in cities" as considered by Francis and Lorimer (2011). Such policies have to be complemented by financial incentives, regulations, and funding and demonstration of research projects. All these are belonging to some of the actors mentioned by Francis and Lorimer (2011). And in these fields does the article build recommendations. An application for Athens, Georgia, US, is proposed.

Ottelé et al. (2011) provided an insight into the problem touched by Francis and Lorimer (2011) of the socioeconomic aspects of green walls. It did a comparative life cycle analysis (constructing, maintaining and disposing) of green facades and living wall systems. The article performs a comparative life cycle analysis for the following:

- a conventional built up European brick facade,
- a facade greened directly,
- a facade greened indirectly (supported by a steel mesh),
- a facade covered with a living wall system based on planter boxes and
- a facade covered with a living wall system based on felt layers

situated in the Netherlands. The environmental impact comprises energy benefits and hints to economical benefits which could be estimated, thanks to durability, aesthetical value and social factor. The direct greening system, the indirect greening system with a supporting system and the living wall system based on planter boxes are the environmentally preferable choice when constructing and retrofitting a building despite being in construction more resource costly.

Ichihara and Cohen (2011) discussed more directly the costs of green space intervention on buildings. Green roofs might have an impact on property values such as rental pricing. In Bostenaru Dan and Mendes (2013), we discussed the way archive research or even games research can be done to see the improvement of green spaces in the vicinity of buildings—the space saved for the green space has the costs saved by the increased property value of the neighbouring parcels according to Woinaroski (2011). The method used by Ichihara and Cohen is regression.

Chang et al. (2010) looked as well to the costs of green roofs, related to the cistern volume. Water conservation and reuse of rain water play a key role.

Conway et al. (2010) looked to the effects of urban green space on residential property values as well. The model used, like in the regression employed previously, is a hedonic model. The case study is this time Los Angeles. Spatial dependence is taken into consideration (a spatial model per se, and spatial lag). The spatial model employs GIS software in which the green cover has been identified. There is a positive impact of green on the property value, which makes greening incentives to be desirable, even funded by property owners. Greening strategies are however not a replacement of park spaces. It is to be noted the green walls can be more difficult to assess through such a spatial approach than the green roofs already investigated, and oblique imagery might be necessary.

Other studies on the costs of green roofs have been performed by Blackhurst et al. (2010) and Houdeshel et al. (2011).

Another article looking generally to green facades is Köhler (2008). It starts with a view back in history and the climbing plants of the nineteenth century, although vines in the Mediteranean region are a 2,000 years old form of vertical gardens, after which it proceeds with the analysis of German language literature. The paradigm shift has been promoted in Germany by the artist Hundertwasser. http://www.garden-cult.de is an open-access literature catalogue documenting the use of green walls since 1880 till 1940s, when the living wall patents were issued, in German language literature. The author completed his PhD as early as 1987, along with two other ones in Germany, on green facades. The article also provides an overview of the benefits and costs of green facades. As ongoing research, the article identifies some in building physics, as we will analyse in the following.

Pérez et al. (2011) looked at the energy savings "green vertical systems" can provide within the approach to a house as passive house. The study only considers exterior green walls. Possible energy upgrades are provided by green facades of the type double-skin or green curtain through shadow, insulation, evaporative cooling and barrier to wind. The climate considered is dry Mediteranean continental. Pérez et al. (2011) classifies green vertical systems into green facades and living walls. Green facades are considered in this context to be extensive green and sub-classified into traditional and double-skin or green curtain (modular trellis, wired or mesh structures), or intensive green, which are perimeter flowerpots. Living walls are considered to be intensive green and in this category fall panels and geotextile felt. We can see that this classification also applies to the examples we gave. Pérez et al. (2011) give instead examples of producers.

Ip et al. (2010) looks at only one aspect of the improvement a green façade can give: the shading performance. The green façade considered is the canopy (climbing plants) called BioShaders, for which a thermal model based on coefficients has been developed considering the UK climate. The parametric research has been conducted through an experimental investigation over two years.

Wong et al. (2010) look at another partial aspect of improvement through green walls: the thermal improvement. Experimental investigation has been conducted in HortPark, Singapore, for 8 different vertical greenery systems. Vertical greenery systems have a cooling effect in a tropical climate, leading to energy savings through cooling, but depend on the type of the greenery system chosen. The article performs a literature review emphasizing the role research in Germany has, where this has made its way into regulations. Thermal benefits highlighted are temperature reduction as well as shading and insulation in the humid climate of Hong Kong. The eight types considered were as follows:

1. Living wall: Modular panel, vertical interface and mixed substrate.
2. Green façade: Modular trellis.
3. Living wall: Grid and modular, vertical interface and mixed substrate.
4. Living wall: Modular panel, vertical interface and inorganic substrate.
5. Living wall: Planter panel, angled interface and green roof substrate.
6. Living wall: Framed miniplanters, horizontal interface and soil substrate.
7. Living wall: Vertical moss-tile, vertical interface and inorganic substrate.
   (a) Living wall: Flexible mat tapestry, horizontal interface and soil substrate.
8. Living wall: Plant cassette, horizontal interface and soil substrate.

It proved that VGS 4 has the best overall performance, while VGS 3 and 4 have the best cooling capacities.

Perini et al. (2011) looked at a similar issue, the effect of vertical greening systems on the airflow and temperature at the building envelope, both in summer (in warm climates) and in winter (in cold climates). Here, also an experimental study has been conducted "to measure the temperature (air and surface) and the airflow near and on different types of green façades and a living wall system", also compared to a bare façade. It is differentiated between green façade, which consists of climbers, attached directly to the building surface traditionally or on trellis or cables (out of different materials, not just steel as mentioned here). Living walls are considered synonymous to green walls or vertical gardens and consist of modular panels which contain the growing medium. The thickness of foliage, air and water content as well as material properties of the panels can influence their properties. For facades in three localities in the Netherlands, the following systems have been considered:

1. a direct façade greening system
2. an indirect façade greening system
3. a living wall system based on planter boxes filled with potting soil.

Results show that "the direct greening system and the living wall system based on planter boxes are the most effective wind barriers", while "the system with the major impact on the thermal resistance is the living wall system based on planter boxes".

Jim and He (2011) looked at the heat flux transmission of the "vertical greenery ecosystem". Experimental research has been conducted in Hong Kong in order to calibrate the developed thermal model through which numerical simulation has been conducted as well. The green wall canopy has transmittance, absorbance and reflectance properties of solar radiation.

Other studies related to the energy aspects are as early as Holm (1989) up to Sheweka and Magdy (2011) or relate to the contribution of the "green" plants to the ecological "green" issue (Luo et al. 2011 as well as Weinmaster, 2009 and Haggag, 2010), while others look at the botanical ground (Salas et al. 2010)

3 Results: Research by Design

We tried to develop some green wall systems which differ from those seen through our field research, or which are rarely investigated in the literature, by connecting them to certain architecture concepts. Such one is first the reconstruction after disasters. Another one is the use of indoor biowalls in dialogue with the living walls integrated in a certain architectural style or as exhibition item. And finally, we will discuss the green façade as a drawing of plants on a media façade, the Art Nouveau of today.

3.1 Green Walls and Propping Systems

Finally, the most important step in our work is the research by design in creating a new type of wall, suitable for the analysed situation. A special support skeleton for the earth and seeds and geotextile module, inspired by the earthquake-resistant

Fig. 22 Vegetation on ruins. Faial, 10 years after the earthquake. *Photo* M. Bostenaru 2008

traditional timber construction, suitable for both indoor and outdoor is in development, with the material savings provided. On this skeleton, not only such a felt module can be brought, but it can serve also as support for climbing plants. We therefore build on the idea behind "green facades" rather than "living walls". It is investigated how this can be brought in agreement with propping systems. Further on, it can be brought in agreement with representations of memory after disasters, with plants growing on the ruins (Fig. 22), which we identified not only at natural disasters, but also at abandonment (Fig. 24).

During a field trip to see the aftermath of the Azores 1998 earthquake 10 years after (Fig. 22), we noticed the ruins reconquered by vegetation as in paintings of the Romantic movement such as the ones by Caspar David Friedrich.

At the Canadian Centre for Architecture (CCA), such green walls on ruins have been however designed on purpose. The ruin of the Shaughnessy house, placed in the garden of the CCA on the opposite side of the road, is a monument in the memory of demolition in this part of the city (Fig. 23). On these ruins, green plants climb.

But sometimes, abandonment of the building is accompanied by abandonment of the plants. Such is a hospital designed by Henrietta Delavrancea-Gibory in Romania (Fig. 24). The plants have dried since the building is deserted.

An abandonment situation is one of the centre of L'Aquila in Italy after the 2009 earthquake. Instead of being repaired ever since, the buildings of the centre have been propped with mixed timber and steel systems to avoid further deterioration (Fig. 25) and then lacking further funds not repaired anymore. We propose to use these support systems which were also inspiration for an installation by Michele de Lucchi (http://www.youtube.com/watch?v=QQT_GUeJZHY) for new

Fig. 23 Artificial ruin in the CCA garden, with view to the real Shaugnessy house to which it refers. *Photo* M. Bostenaru 2010

Fig. 24 Henrietta Delavrancea-Gibory. Abandoned hospital. The green wall is also abandoned. *Photo* M. Bostenaru 2011

Fig. 25 Temporary propping in L'Aquila one year after the earthquake. Inspiration for Michele de Lucchi. *Photo* M. Bostenaru 2010

Fig. 26 Wall systems in the reconstruction after the 1755 Lisbon earthquake. Contemporary intervention on such a Pombalina type house

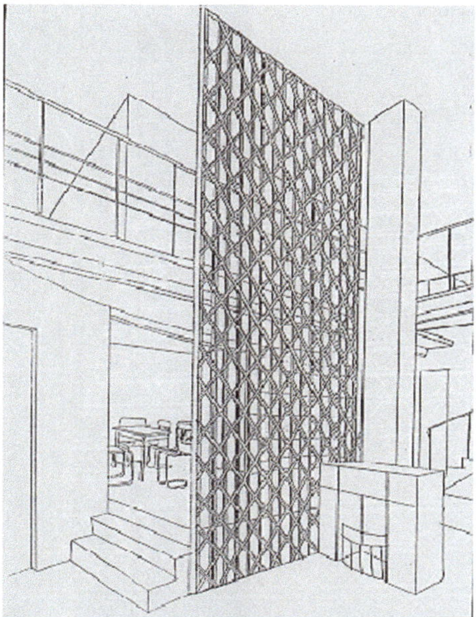

Fig. 27 Timber structure as separation © M. Bostenaru Dan 1993

Fig. 28 Pockets, filled out with felt and water and vitamins, creating green walls. *Photo* M. Bostenaru 2012

Fig. 29 Green wall drawings in Art Nouveau hotel Regency, Milan. *Photo* M. Bostenaru 2011

vegetation to grow on them, either climbing, or in superposed "pots", out of felt pockets. The timber system of the propping reminds of the earthquake reconstruction system in X of the gaiola pombalina in Portugal (Fig. 26) and of a system of supporting such walls which we designed (Fig. 27). Our approach is along with another land art approaches for this abandoned site, for example implemented by the PAO association (http://www.pao-project.org/?p=76) which created temporary structures to revive the public spaces in an experimental architecture summercamp.

Figure 28 shows a green wall pot system with felt to be hanged on such an installation.

Therefore, we propose to make the installation fill a window, if not possible in the building, then windows in the ruin in the CCA garden. The digital version will be then included in the main exhibition, to raise attention. In winter, it is expected that the vegetal cover is dry, like in abandoned buildings such as Henrietta Delavrancea Gibory's hospital canopy in Romania.

3.2 Green Walls and Multimedia Installation

In November 2011, we participated at a meeting on Art Nouveau and Ecology in Milano. In the moderated discussions, it came out that the drawn plants of Art Nouveau might be today's green walls, and, as such, the Art Nouveau style inspired today's ecologic movement. In the plants of the drawn tappets (Fig. 29)

Fig. 30 Green wall drawings in multimedia digital installation of plants growing on a wall in ZKM, Karlsruhe. "interactive plant growing" installation by Christa Sommerer und Laurent Mignonneau 1992. *Photo* M. Bostenaru 2013

and stained glass windows of Art Nouveau, however, any association was possible, while for the green walls of living plants, the specification of the respective location has to be respected in order that the plants live. The autumn of 2013 brings an travelling exhibition of the Art Nouveau and Ecology project called "The Nature of Art Nouveau" to different cities in Europe, starting with Helsinki and Brussels. To Art Nouveau, it was not foreign, however, also the relationship to botany and more widely to the evolution theory and to view of biological items in the microscope (Sander 2002; Gamwell 2003; Martin 2009 and Kolijn 2013).

"The exhibition is divided into three sections:

- The eye of the period, which examines artistic fascination with nature, with scientific discoveries in botany, zoology, oceanography, with evolutionary theories and technical inventions of microscope, photography and cinema.
- The Nature's Workshop, which shows artistic experimentations with natural and artificial materials and new techniques to create new forms, inspired by nature: new line, ornaments and structures, patterns and symbols.
- Artificial landscapes, which illustrate art nouveau interpretations of natural periods and elements—spring, earth, water, air, light, fire." (http://www.artnouveau-net.eu/ArtNouveauEcology/Actions/Exhibitions/tabid/158/language/enGB/Default.aspx)

Today's multimedia techniques allow also other kinds of drawings. It is the multimedia which can connect the living plant and the drawing on the wall, as it was designed in a digital installation already in 1992 at the Centre for Art and Media in Karlsruhe (Zentrum für Kunst- und Medientechnologie - ZKM) (Fig. 30). Here is the plant material dialogue with the drawing in a pioneering work of its kind.

The ZKM showed a constant interest in connecting the display of nature in science, in a modern interpretation of what Art Nouveau did. As such, the Olafur Eliasson exhibition, apart of building installations which display natural phenomena such as a rainbow, also featured the "moss wall" (http://www.olafureliasson.net/works/moss_wall.html) in the exhibition "Surroundings Surrounded" in 2001. The Olafur Eliasson exhibition returned in 2008 to ZKM as "The Spiral View". The "Moss Wall" is an installation from 1994, so somehow contemporary with the other exhibition piece, to bring the way Iceland moss acts on the senses (not only visual, but also smell) to the closed space of a museum (http://www.youtube.com/watch?v=maH4un17gc0). As natural environment it connects to those shown by us at the begin of the paper, where green walls are created naturally in humid environments. As artistic project it can be best connected to our "land art" aim. And also as artistic project it is remarkable the way the memory of senses is being talked at.

In our opinion, the ZKM approach can be the starting point for installation transforming blind walls into multimedia facades with drawn plants like in Art Nouveau, following a current of multimedia facades such as the ones designed also by Dan Hanganu for example the Pointe-à-Callière Archeological Museum in Montreal.

An important aspect of conservation is to preserve the memory of the architecture object, which is why we propose an incursion by a multimedia art project, multimedia playing itself a role in conservation interventions for the relationship between old and new. Such a multimedia incursion was done by Hanganu himself in the lighting at the Archeology Museum, and in other museums, outside Montreal, and is therefore suitable for the "homage". Although multimedia was brought to architecture already in the interwar time (the Fontana del'Essedra at Mostra d'Oltremare in Naples, Italy), and brought to highest expression to also Romania-born Iannis Xenakis, to whom the CCA dedicated an exhibition, we aim to use today's everyday accessible technology to realize the project and thus integrate with current trends of digital methods in the arts.

We stress the role of photography. Our support grant at the CCA was on photography, and after all, we propose an alternative photograph exhibition. Photography has potential for augmented reality in the comparison "then and now". Instead of the classical "then and now" comparison, we compare full colour image projection and shadow. And we propose a participative exhibition, where the visitors are projected onto the space of the images/shadows—another kind of augmented reality.

The virtual preservation through photography of historic buildings raise the question of keeping the memories of the work. For this reason, we undertake a research project, "research by design", on the artistic representation of memory. Initially, we conducted research with methods offered by philosophy, phenomenology starting from Heidegger, phenomenology to which we turn and later, when we will involve participatory architecture. We created, we defined "the rediscovered space" of buildings by Dan Hanganu. Photography in itself is an access "door" to memory.

We present an installation symbolizing a unique encounter scenography. The doors to the rediscovered space merged a long time ago with the environment. We have a door which belongs to the past, until a space yet known was rediscovered, and have a light, the light of today, trying to rediscover, trying to show us a path, or even and what is behind the door, but what we have a glimpse of. When archaeologists are trying to find some signs of a culture that no longer exists, or at least not as it was, they collect some items, and include them in the museum and then say: "You cannot get a sense of culture". The idea that is that very rarely that a museum of cultural attractions really reveals a culture we have built memories of. Therefore, we propose this installation inspired by the description in Michael Ende's authored fairy tale of how to enter another world. What we try is exactly make visitors enter the screen, through a 1:1 simulation of virtual reality, not only through glasses as in stereo views.

Such an installation could be also extended to the city, projecting, for example on city walls an approach already used by cities for festive occasions, a new way of painting, having the art of Art Nouveau instead of planted green walls, as we will show in the other, connected proposal, since Dan Hanganu also designed a building with a green wall. In Bucharest such a media facade could be seen at the Christmas market in 2013.

The shadows, the reflection and the projection of photographs of Hanganu architecture in Montreal will be carried out, while permitting the walk of the public between the two elements of these multimedia doors and thus the integration of the shadow or the mirroring of the person visiting the installation into the landscape of Hanganu's architecture. Like in Michael Ende's "Neverending story", these doors open from just one side, the one enriched by the multimedia element. From the other side, they are opaque. While the roll-ups will be kept vertical, adequate for portrait photography, and rather at the size of a door, the mirror will be horizontal, adequate for landscape orientation photography, proving with this that a door can have any shape. A rather long and shallow space of the CCA, like a corridor, is adequate for this. The shadows will be created through DTP transformation, to create the effect of aluminium foil on traditional slides, to create the effect of projecting a light over a hard copy model. A computer will display on the two screens the shadow and the projection of the same image. All items can be purchased in Canada on the place, and easily installed and uninstalled, being possible that the CCA already has some, or, if not, they will be added to the inventory for daily business, as they are adequate for this.

3.3 Innovative Green Walls as Separation in House Design

Finally, we designed two houses to integrate green walls in the interior and the exterior. While for the exterior, our solutions are inspired by Patrick Blanc's approach, for the interior, we wanted to design something new, namely transparent green walls. For this reason, we lean on the design we explained for the propping systems, where the new wall can fill a window like at Michele de Lucchi. It can be done either out of climbing plants, or of the shown pockets hanged at a certain distance.

The first green wall design in a house is house "Mondrian" (Fig. 31), a residence with an artist's workshop, like in the experiments of the Avant-Garde in the 1920s. Outgoing point was a painting by Piet Mondrian "Tableau No. IV. Lozenge

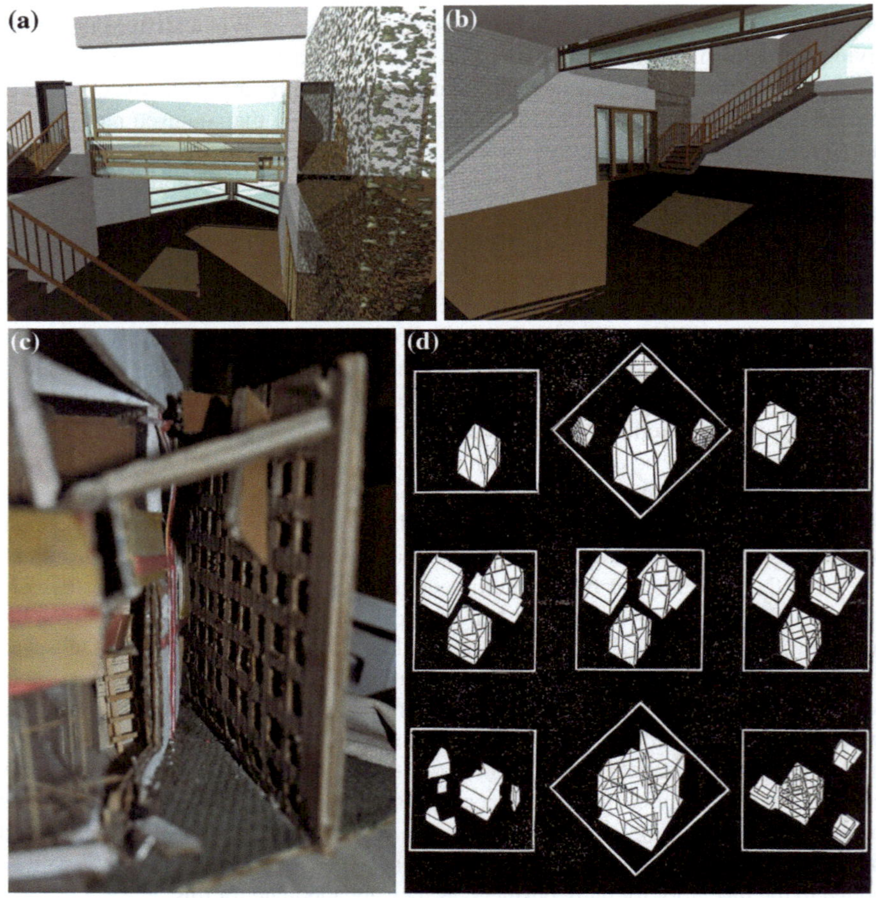

Fig. 31 House Mondrian. Green walls and morphogenesis © Bostenaru 1993–2013. Renderings (**a**, **b**) to show the interior space, part of the model showing the exterior all (**c**) and morphogenesis following the Eisenman example but applied for the space definitions resulting from the Mondrian painting (**d**)

Composition with Red, Grey, Blue, Yellow, and Black, 1924/1925, oil on canvas on hardboard, National Gallery of Art, Washington". The design method was based on morphogenesis. The painting was copied in black and white, thus resulting in grey tones which determine through the intensity of the tone the height of the space. Thus, it results in a building of the type "Raumplan", following the concept of Adolf Loos on the relationship between space and structure. The morphogenesis process can be well presented following the method of Peter Eisenmann at House III, Lakeville/Connecticut (1969–1971), but the same method has been employed by Giuseppe Terragni for Casa Giuliani Frigerio in Como in the 1930s. Recently, Augustin Ioan launched the project "Marcel Janco reloaded" with this approach on 1920s architecture, even with artist's workshops. It seems important

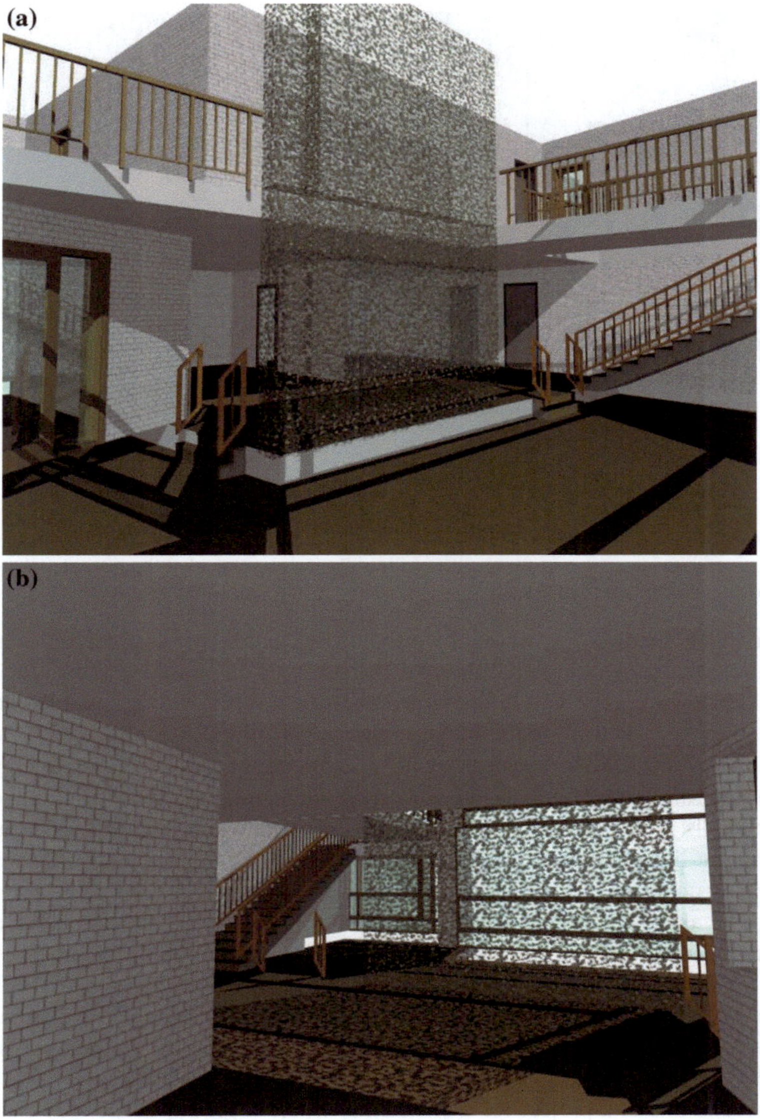

Fig. 32 One-family house with transparent green wall © Bostenaru 1994–2013

to us that the house is a cube. In a research approach called "Rediscovered space", we defined spaces to inspire different feelings through a morphogenesis process from 2D to 3D in which this space is defined as interior of a cube. Our cube is more complex, having also a function. Going back to the inspiration from Mondrian we remark the relationship which we have to the Neoplasticism movement in the 1920s, which inspired the De Stijl style in architecture through

Fig. 33 Project for a water museum in Frankfurt on the Maine, the central hall. The organic timber beams can serve as support for a green ceiling. Project developed at the institute where Alex Dill, co-author in the other contribution, was academic councillor © Bostenaru 1999

Mondrian like paintings. Although we took Mondrian's painting, we did not consider the colours. The brown colour of the earth in the model was to life in our project through transformation into green walls, and this is the sustainable concept. We have an interior wall, in the initial project a perforated concrete wall, on which plants can climb. On the facades, we have green areas, to be realized through the felt concept developed by Patrick Blank. And in the garden, we have again the perforated walls which become green and which come close to the glass façade of the building like at Stilwerk, Patrick Blanc's work in Vienna. Finally, Mondrian's painting is characterized through rotation at 45°. We took over this rotation into the façade, and through this, we step in relationship with Italian Rationalist architecture. Diagonals are a way of protection against earthquakes, and an approach with coloured diagonals on white background was performed in the south Italian province of Potenza at the retrofit of Scuola Domiziano Viola in 2010. The seismic approach and the presence of the environmental element of the green walls contribute to the sustainability of a building with an original concept.

A second one is a project for a one-family house (Fig. 32) in the way designed after 1989 in Romania, with generous spaces, a living room on two levels, staying alone. The plan layout is based on the curve line which crosses the plan, the straight elements being rays towards the centre of this circle. The curved line is represented by a glassed wall, through windows at the exterior and through glass brick on the kitchen wall at the interior. The green wall, the sustainability element of the residence, marks, as in the other project, the inflexion point of the staircase. The stair is mainly in one flight, with just several steps on the second one, followed by a stage in the night zone on the upper floor. On the ground floor, the green wall has the role to separate the dining place from the living space, situated few steps down. On the upper floor, it separates the playground of the children from the empty space over the living. The green wall has been designed to be transparent. It is based on the growth of plants in a similar way to that in Dublin on the timber trafor instead of steel and was designed, as we saw, specially for this house with material economy.

With this concept of transparent separation walls, the proposal developed by us draws on the transparence Art Nouveau promoted by drawing plants on stained glass.

A further project of employing green walls follows the idea of suspended walls for an intervention within the Green Belt of Frankfurt on the Maine, Germany, with a building of Aquarium and Museum of Water (Fig. 33) in the Old Port, which emphasized the two-fold characteristic of water, giving life and being flood hazard. Like at the Aquarium in Genova for which we showed the Patrick Blanc wall, it is again a way of expressing the blue-green infrastructure at the level of the building, which even is conceived as museum for this at a suitable site. Such developments can be coupled with recent developments in green wall building to build symbolic spaces for protection from hydrologic hazards, or as memorials after hazards. Green should be the ceilings, this time the interior roofs over the hall of the building, between the aquaria, and also green the slopes of the outside garden. The technology of green ceilings has been developed by Patrik Blanck as well. This emphasizes also a characteristic of green walls we talked about at the beginning, which is the connection between water and plants, instead of earth and plants. (Fig. 33).

4 Discussion

With this research, we followed several objectives:

- Improve resilience to climate change by extending the green spaces in the city, by improving understanding on the impact of intensive construction, by developing green plans within the urbanism plans, especially in case of urban restructuration plans.
- Diversification of the application domains of the felt geotextile products, as well as diversification, according to identified new domains, of the product, through the way they are put in oevre, as well as the plants allowed. Some such are already identified and explained.
- Develop sustainable emergency housing models which might be implemented in case of reconstruction after a catastrophe. Inclined roofs are especially suitable for this, but also interior walls.
- Development of a green wall system to support the creation of pocket parks in place of collapsed buildings in catastrophic events through the restructuration. For such a wall system, it will be necessary to use another textile, not biodegradable. A support system out of wood (which is a more durable material), unlike steel in other countries and with more economic layout, will be developed.
- Highlighting the comprehensibility of the measures analysed, through inclusion into the integral planning scheme near the flow in the physical implementation of the green spaces also of a flow for the education of the population which

has to support the measures. This is especially important in the so-called second phase of strategic implementation, i.e. when the application of the measure has to spread from pilot or demonstrative projects to "routine" wide ones.

- Investigation of the possibilities to support such changes by political and economic environment, namely of existing programmes and their impact.

5 Conclusions

"The planner of today cannot go out from a "tabula rasa" situation any more. Environmental and sustainability issues have already formed the public idea that a "green belt" of our cities is necessary, a kind of fortification leading to intensive development of towns inside a clearly delimitated area within the surrounding nature. Since building on the periphery is limited, and the existing built substance has a certain cultural, architectural or at least environmental value, upgrading of existing buildings gains more and more ground from the design of new buildings". (http://www.roseschool.it/ca_redivivus/ text by the author) While in the majority of cases, this upgrading means bringing the buildings to the comfort required by changed living standards, we focused in this paper on the city-nature dialogue by upgrading building with transforming their vertical surfaces in vertical green surfaces of plants.

In this paper, we investigated how nature, separated from the densely built city, if shrinking or growing to be revalued within its limits, can be brought back to this. Current environmental problems as we face them because of climate change are encouraging to have green spaces in the city. But space is scarce, hence the need to recycle unused potential, such as empty parcels for pocket parks and empty walls for the so-called vertical garden and its alternatives. Healthy spaces with green have been already promoted by the functionalist movement, but as replacement for the traditional building blocks. Such vertical green surfaces use exactly the characteristics of building blocks for implementing them and are as such an intervention for the conservation of the urban tissue. We will see how patches of such green spaces can be created in the approach in Chap. 12.

In the meantime, the mentioned issue of changed living standards may not be neglected. The vertical green has to be brought in accordance with habits of the population; it has to be accepted by them. We found in the reviewed literature a study on the perception of beauty in this regard, but these studies may be extended.

However, the vertical garden is expensive and so far only show-off projects have implemented it. It is the reason why it is less spread, although spectacular, than green roofs, which are included in the policies of some cities. Maintenance is also a problem, as the issue of irrigation has to be solved satisfactorily for the green walls to look well. And finally, the climate is an issue. Humid warm climates are more suitable for this kind of walls, since there vegetation is also growing naturally on stone walls. For this reason, apart from the felt-based vertical garden, also alternative solutions have been looked for. Some of them are artistic projects like those of Hundertwasser, or the approach by Eliasson which builds on the natural occurence of green (moss) walls in humid areas which also inspired Patrick Blanc also some show-off projects, but some of them go back to the history of vertical surfaces as support for depicting vegetation from Art Nouveau and depict them now by multimedia means. But most suitable for further research and employment are the superposed pots. Research has been done to classify interventions for vertical green surfaces, and we categorized the case studies we looked at according to these.

The studies we reviewed demonstrate that green spaces enhance the value of adjacent parcels so far that these do not have to be occupied by a building to come to the same value of the whole complex. This is also demonstrated by simulations of the socioeconomic model in games. However, we still miss studies to compare the different kinds of green walls, living walls or other, as we have for energy studies, in building, and in maintenance. These are future construction management research projects we aim for. Best practices can be based on these issues of more likelihood of spreading in implementation or in best maintenance and thus showing the potential. Pocket Parks are usually paid for by the community around, but in some countries like Romania by the public administration. A similar model could be done for the exterior green surfaces of the buildings. Apart from the need of being maintained, green vertical surfaces contribute to the maintenance of buildings. They help saving energy costs, and recent developments of building information modelling allow including such modules into software like archiCAD.

Although the vertical garden on felt-based surfaces is currently a trademark of the French botanist Patrick Blanc, the patents have been released in the interwar time in the US, but, given the problems of expenses depicted above, not employed. A longer history have instead climbing plants, for which the literature ranging from the nineteenth century exists. Such have been employed even in interwar time, when architects looked for new technological materials and separated from nature, even in Hundertwasser manner, and up to today in works such as of architect Dan Hanganu.

Usually, urban planning is seen separated from interior design. However, here, with the contribution of landscape, we have a dialogue between the detail in the interior of the house and the large scale of the city centre separated from the rural and agricultural area around, the countryside. We tried more than bringing nature back to the city centre. We brought it to the interior of the house.

The review of case studies and of the literature served us to make something typical for architectural research, which is research by design. We first designed installations based on the artistic approach in Art Nouveau and at the Centre for Arts and Media (ZKM) and then two houses which feature green walls inside and for one of them also outside. They dialogue with the interwar architecture language, bringing modernism to a response today to current challenges of climate change. The approach in our own projects is innovative, since it breaks the opaque wall and makes it transparent, like Art Nouveau stained glass. For such an approach, either the superposed pots or the climbing plants are suitable, so either the traditional or the alternative to the felt-based wall. Such interior walls are also more protected from climate, as several examples, not reviewed in this contribution, show. If we were to connect to the view of Art Nouveau, the development of plants as fractals could link again to the multimedia view, but also to the view how we see a grid network on which they could grow we show in Chap. 12 and thus improve the support structure designed here.

Acknowledgments This paper is based on work developed within a DOMUS scholarship supported by the Hungarian Academy of Sciences at the Corvinus University of Budapest. The views and conclusions contained in the paper are those of the author and do not necessarily represent those of the Hungarian Academy of Sciences.

References

Blackhurst MPE, Hendrickson C, Matthews HS (2010) Cost-effectiveness of green roofs. J Archit Eng 16(4):136–143

Bostenaru Dan M (2004) Multi-criteria decision model for retrofitting existing buildings, Natural Hazards and Earth System Sciences 4:485–499

Bostenaru Dan M, Mendes D (2013) Contemporary installations with timber from disaster reconstructions. In: Correia M, Carlos G, Rocha S (eds) Vernacular heritage and earthen architecture: contributions for sustainable development. CRC Press, Zug

Carter T, Fowler L (2008) Establishing green roof infrastructure through environmental policy instruments. Environ Manage 42:151–164

Chang NB, Rivera BJ, Wanielista M (2010) Cost benefit optimization of Cistern volume and green roof area in the Florida showcase green envirohome (FSGE). In: Proceedings of the 2010 IEEE international symposium on sustainable systems and technology, ISSST 2010, art. no. 5507717

Colding J (2007) 'Ecological land-use complementation' for building resilience in urban ecosystems. Landscape Urban Plann 81(1–2):46–55

Conway D, Li CQ, Wolch J, Kahle C, Jerrett M (2010) A spatial autocorrelation approach for examining the effects of urban greenspace on residential property values. J Real Estate Finan Econ 41:150–169

Deleuze G (1980) Mille plateaux. Minuit, Paris

Dvorak B, Volder A (2010) Green roof vegetation for North American ecoregions: a literature review. Landscape Urban Plan 96(4):197–213

El Adli Imam KZ (2006) Role of urban greenway systems in planning residential communities: a case study from Egypt. Landscape Urban Plan 76:192–209

Francis RA, Lorimer J (2011) Urban reconciliation ecology: the potential of living roofs and walls. J Environ Manage 92(6):1429–1437

Gamwell L (2003) Perceptions of science beyond the visible–microscopy, nature, and art. Science 299(5603):49–50

Haggag MA (2010) The use of green walls in sustainable urban context: with reference to Dubai, UAE. WIT Trans Ecol Environ 128:261–270

Henry A, Frascaria-Lacoste N (2012) The green roof dilemma—discussion of Francis and Lorimer (2011). J Environ Manage 104:91–92

Hindle RL (2012) A vertical garden: origins of the vegetation-bearing architectonic structure and system (1938). Stud Hist Gardens Des Landscapes: Int Q 32(2):99–110

Holm D (1989) Thermal improvement by means of leaf cover on external walls–a simulation model. Energy Build 14(1):19–30

Houdeshel CD, Pomeroy CA, Hair L, Moeller J (2011) Cost-estimating tools for low-impact development best management practices: challenges, limitations, and implications. J Irrig Drainage Eng 137:183–189 SPECIAL ISSUE: Urban Storm-Water Management in the 21st Century

Ichihara K, Cohen JP (2011) New York City property values: what is the impact of green roofs on rental pricing? Lett Spat Res Sci 4(1):21–30

Ip K, Lam M, Miller A (2010) Shading performance of a vertical deciduous climbing plant canopy. Build Environ 45(1):81–88

Jim CY, He H (2011) Estimating heat flux transmission of vertical greenery ecosystem. Ecol Eng 37(8):1112–1122

Joedicke J (1976) Angewandte Entwurfsmethodik fur Architekten Applied Design Methodology for Architects in German Karl Kramer, Stuttgart

Koenigs T (ed) (1991) Vision offener Grunräume: GrünGürtel Frankfurt = Vision on Open Green Spaces: GreenBelt Frankfurt (in German). Campus-Verl, Frankfurt/Main

Köhler M (2008) Green facades—a view back and some visions. Urban Ecoyst 11(4):423–436

Kolijn E (2013) Observation and visualization: reflections on the relationship between science, visual arts, and the evolution of the scientific image. Antonie Van Leeuwenhoek 104:597–608

Kraftl P (2009) Living in an artwork: the extraordinary geographies of the Hundertwasser-Haus, Vienna. Cultural Geographies 16:111–134

Lundholm JT, Peck SW (2008) Introduction: frontiers of green roof ecology. Urban Ecosyst 11(4):335–337

Luo Q, Li N, Shi MH, Liu Q (2011) Green construction evaluation of wall project based on grey clustering method. Adv Mater Res 243–249:6971–6975

Martin C (2009) Evolution's influence on art nouveau. Nature 460:37

Molineux CJ, Fentiman CH, Gange AC (2009) Characterising alternative recycled waste materials for use as green roof growing media in the U.K. Ecol Eng 35(10):1507–1513

Ottelé M, Perini K, Fraaij ALA et al (2011) Comparative life cycle analysis for green façades and living wall systems. Energy Build 43(12):3419–3429

Pérez G, Rincón L, Vila A, González JM, Cabeza LF (2011) Green vertical systems for buildings as passive systems for energy savings. Appl Energy 88(12):4854–4859

Perini K, Ottelé M, Fraaij ALA et al (2011) Vertical greening systems and the effect on air flow and temperature on the building envelope. Build Environ 46(11):2287–2294

Salas MC, Montero JL, Moral JA (2010) Hydroponic system for growing ground cover plants on vertical surface. Acta Horticulturae 881:421–424

Sander K (2002) Ernst Haeckel's ontogenetic recapitulation: irritation and incentive from 1866 to our time. Ann Anat: Anatomischer Anz 184(6):523–533

Scherba A, Sailor DJ, Rosenstiel TN, Wamser CC (2011) Modeling impacts of roof reflectivity, integrated photovoltaic panels and green roof systems on sensible heat flux into the urban environment. Build Environ 46(12):2542–2551

Semper G (1863) Der Stil in den technischen und tektonischen Künsten oder praktische Ästhetik, München. http://digi.ub.uni-heidelberg.de/diglit/semper1863

Sheweka S, Magdy N (2011) The living walls as an approach for a healthy urban environment. Energy Procedia 6:592–599

Simmons MT, Gardiner B, Windhager S, Tinsley J (2008) Green roofs are not created equal: the hydrologic and thermal performance of six different extensive green roofs and reflective and non-reflective roofs in a sub-tropical climate. Urban Ecosyst 11(4):339–348

Tachibana D, Naoki S, Maki T, Sato Y, Kikuchi S, Imai K (2011) Important issues in green wall planning necessary to create the fine greening of urban areas. AIJ J Technol Des 17(36):699–702

Wang X, Wu L (2011) Exploration on the renovation of building facades form combined with greenery. In: Proceedings of international conference on electric technology and civil engineering, ICETCE 2011, art. no. 5774420, pp 5989–5992

Weinmaster M (2009) Are green walls as "green" as they look? an introduction to the various technologies and ecological benefits of green walls. J Green Build 4(4):3–18

White EV, Gatersleben B (2011) Greenery on residential buildings: does it affect preferences and perceptions of beauty? J Environ Psychol 31(1):89–98

Woinaroski C (2013) Istorie urbană. Lotizarea şi Parcul Ioanid din Bucureşti în context european, Simetria, Bucharest

Wong NH, Tan AYK, Chen Y, Sekar K, Tan PY, Chan D, Chiang K, Wong NC (2010) Thermal evaluation of vertical greenery systems for building walls. Build Environ 45(3):663–672

Xuesong W and Lei W (2011) Exploration on the Renovation of Building Façades Form---Combined with Greenery, Electric Technology and Civil Engineering (ICETCE), International Conference, Lushan, 5989–5992, doi:10.1109/ICETCE.2011.5774420

Part VII
Urban Landscape: Revitalisation/Itinerary/ Perception—Route

Spatial Street Network and Urban Routes Around the Modernist Boulevard in Bucharest

Maria Bostenaru Dan and Alex Dill

Abstract In January/February 2013, in frame of a short-term visit funded by NeDiMAH at the Karlsruhe Institute of Technology, we returned to an older research on the Magheru Boulevard in Bucharest, shaped during the interwar time (1930s) with Modernist buildings. The main outcome of our research concerned the digital representation of urban cultural routes. For this purpose, we designed some such routes for a number of architects, but also for the green spaces. The information about the architects will flow into an online encyclopaedia and thus be a part of a semantic network. The buildings themselves can be detailed in the plan or in the 3D model. In the meantime, the pedestrian route itself can be connected to the street network. We also analysed the street network as a graph, for two purposes. One is the optimisation of the accessibility through Space Syntax for performing the route. Such results can be used for creating similar routes of pre-1755 earthquake Lisbon in another project of us of similar identification of highlights in frame of a 3D city model. Another one is the optimisation of intervention in case of post-earthquake intervention (evacuation, employment of search and rescue teams) since these Modernist buildings are most vulnerable to earthquakes. For the later, we reviewed literature on the topic including agent-based modelling and street network modelling in GIS. Urban intervention in case of earthquakes can be of interest also for semantic networks, since an ontology of it can be developed.

Keywords Interwar • Route • Fractals • 3D • Space • Morphology • Public participation • GIS • Data • Google Earth • Vulnerability • Space syntax • Perception • Sequential

---

M. Bostenaru Dan (✉)
Urban and Landscape Design Department and Centre for Architectural and Urban Studies, "Ion Mincu" University of Architecture and Urbanism, Academiei 18–20, 010014 Bucharest, Romania
e-mail: maria.bostenaru@iaim.ro

A. Dill
Institute for Architectural Design, Art und Theory, Karlsruhe Institute of Technology, Englerstraße 7, D-76131, Karlsruhe, Germany

C. Crăciun and M. Bostenaru Dan (eds.), *Planning and Designing Sustainable and Resilient Landscapes*, Springer Geography, DOI: 10.1007/978-94-017-8536-5_12,

1 Introduction

Aim of the project was to introduce the Romanian Modernist heritage by means of digital methods to the German community which organises since 10 years a conference to its preservation (see report at Bostenaru 2011a). In 2000, when the applicant was working at the host institution for this Short Visit Grant, a survey of an area comprising 1,500 buildings in the centre of Bucharest was performed and the data introduced in a GIS system. This area includes the main N–S boulevard of Bucharest, a unique section in Europe being built in the interwar time. Today, the boulevard is facing conservation problems, in the context of thermal isolation and seismic retrofit, an example recently altered this way being a multi-storey building by German architect Rudolf Fränkel, who immigrated to Romania that time. The KIT disposes about an archive of this and other German architects in exile. Contacts were done to access archives of other Modernist architects which might be at the family of the emigrated architect Richard Bordenache in Karlsruhe, some buildings of which are in the investigated area. The intervention problems in this context were subject of the doctorate of the first author (Bostenaru Dan 2012). During the work in Karlsruhe those years, the cultural heritage value of the buildings surveyed was not taken into consideration in the digital editing, and this was the aim of the visit. The KIT disposes of the software to introduce these data (including architect, monument value, urban planning protected zone value) in frame of the geoinnovation planning network (http://geo-innovation.stqp.uni-karlsruhe.de/). Aim was also to include the typology into the digital taxonomy overview of the Global Earthquake Model, with corresponding glossary entries, where KIT is a partner. A report on this has been published already (Gallagher et al. 2013). Alternative database applications for the photographic and textual data on these buildings for the Web (e.g. for the DOCOMOMO registries, Sharp and Cooke 2000 where Romania is not a member, but contributed to the UIA project on the topic http://www.archi.fr/UIA/), were to be discussed. Selected architect's biographies will be written and included in the Routledge Encyclopedia of Modernism (http://remodernism.wikispaces.com/; an encyclopaedia being understood as another form of thesaurus and ontology), where we were invited. Some others will build the subject of subsequent urban routes applications. All these different ways of entering information contribute to mapping of the use of digital methods in Europe.

2 Typological Classification of the Studied Type

The typology of buildings itself was studied not only for Romania, but also for numerous European countries, including Italy, Greece and Portugal. The typology was compared to earlier reinforced concrete buildings, with Hennebique system, in order to see to which amount these can belong to structural heritage and to shape the criteria of the architect/conservator.

2.1 The Reinforced Concrete Skeleton Structure in Europe

This research is about the "interior" of multi-storey interwar buildings of European architects, a topic less investigated than that of the facades. Facades, with their elements of style, have already built the subject of numerous studies in history of architecture. The interior is first of all seen at the 2D level of plans, but this functional division, and the grouping of functional units to zones, is experienced in the parcour of the visitor as 3D spatiality. For this, studies of the early reinforced concrete typology in seismic countries, such as Romania, Italy, Greece, Slovenia and Portugal, were conducted, for the first two including archives.

The buildings of the modern avant-garde were raised during a very short time span of 20, sometimes just 10 years, in several different parallelly coexisting styles, coexisting also with the newest development in music, arts, physics, philosophy, economic and social theory and industrialisation. One of the nuclei of the movement was built by the housing programme, particularly suitable to foster innovation. While in more industrialised countries, ways to solve social problems were sought for, and in the other European countries, the new possibilities were seen as an opportunity to give a more prosperous image to cities, by raising density with a block of flats for the middle class.

The site selection subject of the study was enlarged as compared to the initial proposal of the CA'REDIVIVUS project (http://www.roseschool.it/ca_redivivus/). Core countries were Romania, Greece, Portugal, Italy and Slovenia. But it included also countries of mainstream Modernism, such as Germany, the Netherlands and France, the 'Hof's of Vienna, Austria (they are also called Viennaise 'Hof's) and less documented countries such as Estonia, Bulgaria and also forerunners of Modernism in Hungary and Finland. The Central European context, mainly the Czech Republic, was also documented. All these documentations were done based on both literature and field trips. The field trips led to the traces we talk about in this work. This was enriched, when possible, visiting exhibitions displaying archive research results as well as visiting the buildings from inside to experience the spatiality resulting from their floor plans, which was the subject of the project and will be further detailed.

2.2 Related Typologies in Europe

Buildings considered for this study were selected based on the typology: multi-storey structure of reinforced concrete frame. In most cases, they were residential buildings, housing being the predominant programme of the Modern Movement, translated into representatives of the so-called other modernisms. Examples of conservation of residential buildings, however, are rare, so the typology will be extended to other programmes and/or time periods. We find representatives in the countries where residential property is in central urban context, sustainable housing types being identified as such by Sonne (2009). We have chosen in addition to Romania, subject of the study, and Italy, subject of primary comparison, comparative examples in countries with medium or moderate seismic risk moderate. Greece is representative, where the Athens Charter was launched in 1933, defining the rules that have helped define urban typology in Romania, and Portugal, where seismic codes were applied to the frame structure of the early modern period when this was done in timber. Since these are not geographically neighbouring countries, sought to influence the typical structures of other European countries is the mobility of architects.

Multi-stage urban construction had resonance: once increasing the height, to meet hygiene, promoted among others by the avant-garde, then increasing distance between the fronts, and thus, new boulevards were laid out, as in Bucharest, where it was not built in periphery, in the wide green spaces, as in Western Europe. Drawing new avenues superimposed on historic streets lead to irregular contour plots, then reflected in an irregular arrangement of the structural partitioning of the buildings, cause of their seismic vulnerability in Bucharest. In other countries such as Greece and Portugal, boulevards were drawn to extend the city, for example to infrastructure elements such as the train station, and such buildings have a regular structure.

The architecture has led to irregular and functional zoning in the floor plan. Another element of vulnerability is the flexible ground floor, a result of searches to find an expression for the expression of the material of concrete, this time the private public relationship.

Congrès Internationaux d'Architecture Moderne (International Congresses of Modern Architecture—CIAM) started in 1928. The Fourth Congress, where the Athens Chater was proclaimed, was held in Athens in 1933. The book was published only in 1943 by Le Corbusier. The main content of the Athens Charter refers to functional zoning: four independent functional areas: residential, work, leisure and circulation. The Athens Charter has found application mainly in the post-war period, in the reconstruction of cities. In the 1960s, the results obtained in the Western world begun to be criticised, and promotion of alternative progressive urban planning, from strategic planning to participatism, was promoted. In Romania, however, it was effective in some stipulations of the Master Plan of 1934, particularly in defining the architectural volume and functions in accordance with the role of these thoroughfares. In Greece, where Congress was held, the Charter was not implemented. Interwar architectural typology of buildings from different European countries will be analysed in terms of the stipulations offered by CIAM IV.

2.2.1 Building Typology: Romania

In Romania, the interwar architecture was represented by mainly the housing programme, and in the interwar time, the impact of apartment buildings is bigger than of any other housing. A strong economy after the unification process encouraged the private enterprise. Deviations from mainstream international movement were

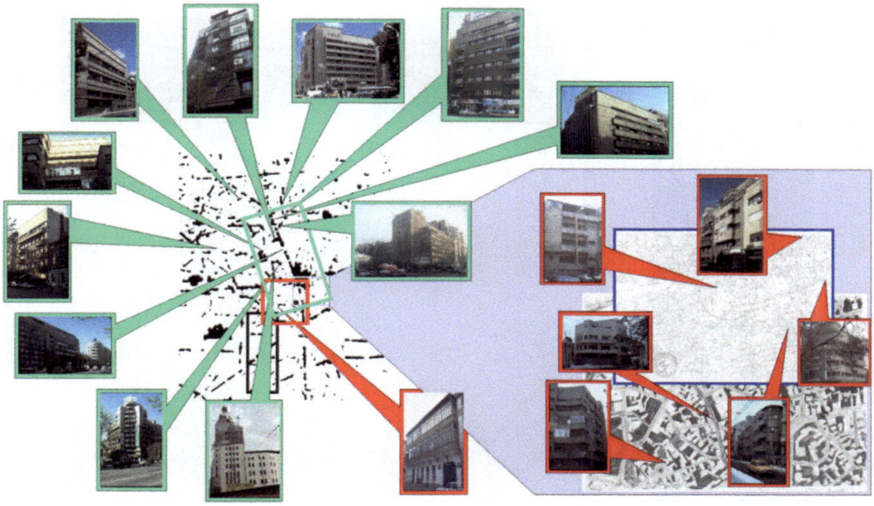

Fig. 1 Early reinforced concrete buildings in Romania. Connection between the test zone on Magheru Boulevard in the project in Karlsruhe and urban routes (after Bostenaru 2006)

dictated by the market (Machedon and Scoffham 1999). These included building in the context in the centre of the city rather than Siedlungen on the periphery (Fig. 1). Nevertheless, the principles of the 1933 Athens Charter were followed in the 1934 Master Plan, and this set up some of the vulnerability issues: the setbacks to keep a constant cornice line, together with higher corner tower developments (some of these towers were cut following damage in the 1977 earthquake) as well as the presence of commercial ground floors, with soft storeys given transparence of the ground floor on columns or even through hall-type functions such as cinemas led to irregular sections, while interior courtyards to win more light while occupying the whole parcel led to irregular floor layouts. The placement of high-rise buildings in the centre, close to the River Dâmbovit,a located these on soft soil, as well as their placement which led to irregular plan layout because of parcels resultingwhen the N–S boulevard was cut is rendering vulnerability. This area in which these buildings are concentrated was extensively investigated in the study. Most buildings that collapsed in 1977 were corned buildings, but there was an exception, the ARO (Patria) building by Horia Creangă, which resisted because it had been shortly before retrofitted to resist to fire.

In Romania, the ownership type in apartment blocks is condominium, like in Greece, until today. Apartments had the benefit of double entrance, which has potential of restoration, with changing family structure. Remarkable is the Elena Ottulescu property (architect Horia Creanga, 1934–1935): an open plan apartment, which is unusual for multi-storey housing. For the structural simulation, the modelling was done for a more regular layout than the real one.

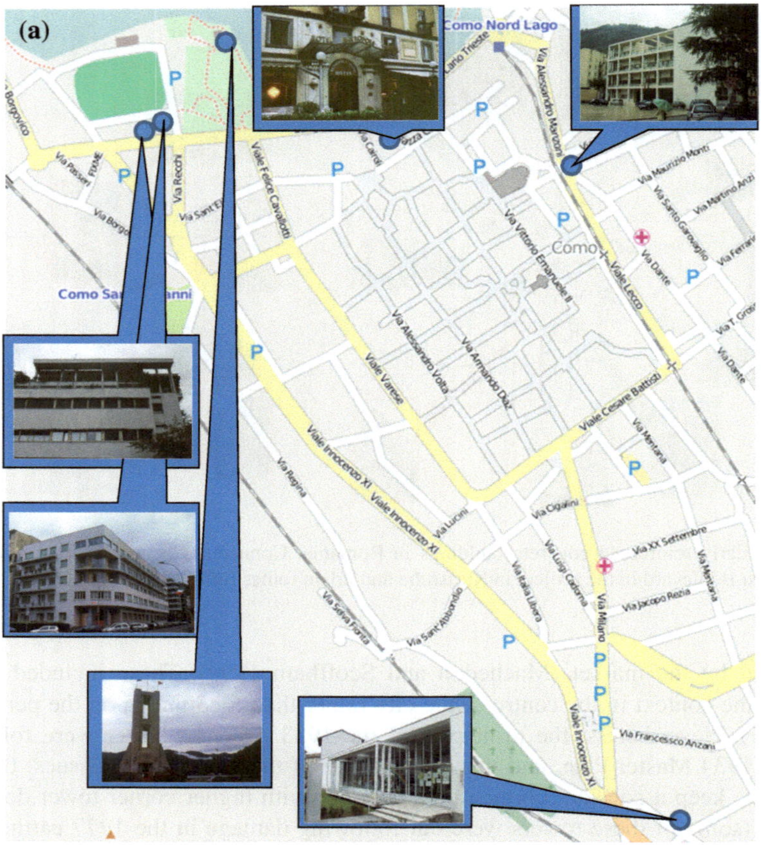

Fig. 2 Giuseppe Terragni rationalist architecture in Como (**a**) and Milan (**b**), photos M. Bostenaru 2005, distribution in the city

2.2.2 Building Typology: Italy

In interwar, Italy architecture followed two directions:

- Rationalism (contextual Modernism), represented by Gruppo 7 whose main representative was Giuseppe Terragni (Fig. 2), but also by other architects such as Adalberto Libera,
- Novecento (Fig. 3)
  - Decorative.
  - Geometrical.

Both Novecento directions were introduced by Giovanni Muzio. For the Decorative Novecento, the manifesto was "Ca'Brutta" (Ugly House) (1922) in Milan, while for the geometric one, it was Casa Bonaiti (1935–1936) at the same location.

Fig. 2 (continued)

Although we mainly analysed these movements in Milan, they were represented also in other cities (Mostra d'Oltremare in Naples, a seismic-prone zone, is an example of rationalism, and the movements can be found also in Rome and other places.)

In Novecento, we have function-bound housing typologies and condominium property. In the zoning, there are function groups, double entrance. While Rationalism was more innovative in the formal language of volumes and facades, in Novecento, the functional organisation and technical solutions, with reinforced concrete which hardly needs to be restored, were innovative. A special case is a villa by Giuseppe de Finetti, an example of stapled villa following Adolf Loos' Raumplan like the Ottulescu building of Horia Creangă in Romania.

As we see from the distribution in Figs. 2 and 3, the buildings of the interwar time in Milan did not form a boulevard like in Romania, and they were spread throughout the city. However, we see such a boulevard in Genova, the Via XX Settembre, with buildings in reinforced concrete but from the turn-of-the-century time and even more resistant to earthquakes because of the geometry of the layout of secondary beams in the Hennebique system. Also, the interwar reinforced concrete buildings of the Novecento are well resistant, due to the high quality of the work.

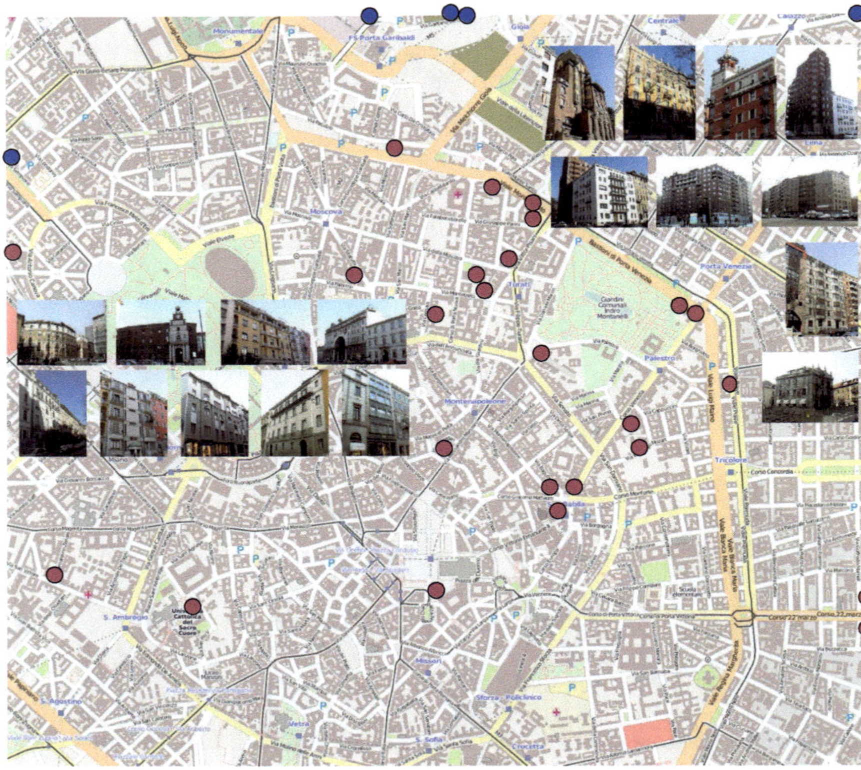

Fig. 3 Novecento architecture in Milan: spread in the city

2.2.3 Building Typology: Greece

In Greece, in 1929, a new law allowed for an ownership system for multi-storey apartments. Housing in private hand seems to be unique (Condaratos and Wang 1999), but actually similar to Romania and Portugal. Architects were mainly receiving training in Germany, little in France, which was the case for Romania. An outstanding example of zonation is the Zaimi and Stournary street example which "resembles Italian rationalism" (Condaratos and Wang 1999). Also here apartments had the benefit of double entrance, for owners and servants. The development in Greece was done in the area extending towards the train stations (Fig. 4), where new streets and therefore with regular parcels were cut block by block. This led to regular floor layouts. There are no buildings with special corner solution developments like in Romania, and the buildings are 6 storeys high at most, like in the Tony Garnier city. The apartment blocks in Athens are not listed as monuments, nor are they currently seismically retrofitted, as it

Fig. 4 Early reinforced concrete architecture in Athens. Photos by M. Bostenaru, spread in the city

is the case for the Army Pension Fund building, an office building converted into shopping centre (Penelis et al. 2003).

2.2.4 Building Typology: Slovenia

Few reinforced concrete skeleton multi-family housing can be found in Ljubljana, as they are mostly with brick structure. Joze Plecnik built housing programmes, but the landmarks created by him in the Slovene capital are rather

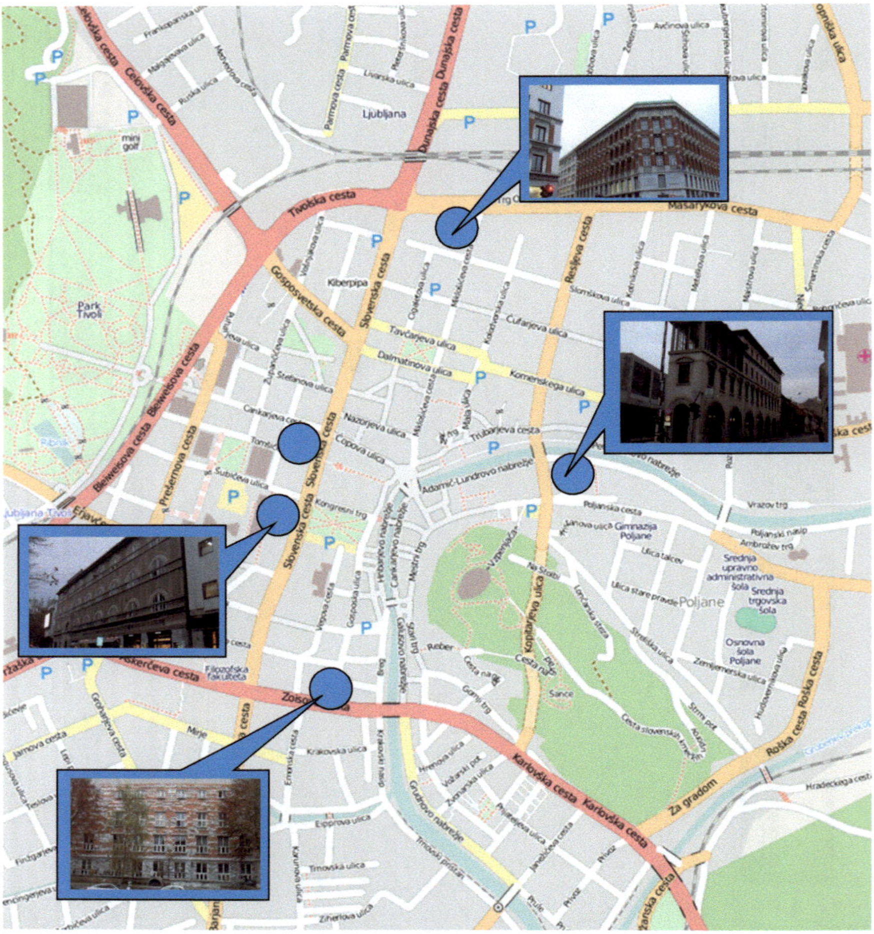

Fig. 5 Joze Plecniki architecture in Slovenia—spread in the city. Photos M. Bostenaru 2008

dedicated to other functions (Fig. 5). Unusual for interwar time, Plecnik also made restoration interventions on ancient monuments. Unlike the three countries analysed till now, in Slovenia, housing programme did not predominate and multi-family housing was mainly in brick. Ljubljana, the Slovene capital, was reconstructed after the 1895 earthquake, mainly with buildings of Art Nouveau; Modernism and reinforced concrete came later. Cultural nationalism was manifested in the choice of the architect to do the then Master Plan. The multi-family housing in reinforced concrete by Plecnik can be found in Vienna (e.g. Zacherl house, which is related in its language to the ARO apartment block by Horia Creangă in Bucharest). Also, in Vienna, there is the first church with reinforced concrete structure, the language of which might have influenced multi-storey construction (Bostenaru Dan 2005a).

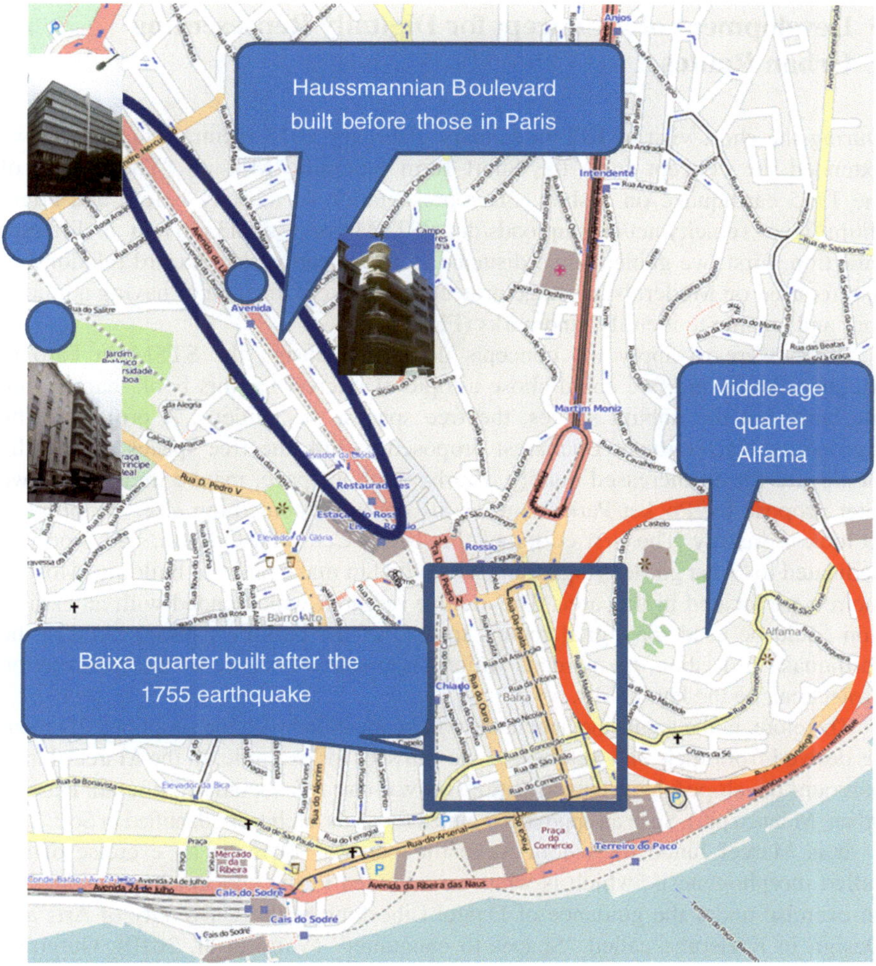

Fig. 6 Early twentieth century development in Portugal: buildings by Cassiano Branco (photos M. Bostenaru 2005) and relationship of the spread in the city to the developments after the 1755 earthquake

2.2.5 Building Typology: Portugal

Reinforced concrete buildings are in the north of the city (Fig. 6), where avenues were built in the interwar time. The Master Plan according to the 1933 Charter of Athens was done post-war (Marat-Mendes and Sampayo 2010). The building follows rather traditional floor plans. Architects of the time were Cassiano Branco, Porfirio Pardal Monteiro and Luis Cristino da Silva, but also Cottinelli Telmo, Rodrigues Lima and Carlos Ramos who built, like also in Romania and Spain, buildings dedicated to cinemas, but also railway stations and apartment block complexes.

3 Development of a Concept for Digitally Representing Urban Routes

During the short visit at KIT in January–February 2013 (central Bucharest), we extended the concept from the Short-Term Scientific Mission "The impact of the 1755 earthquake on Lisbon: 3D representation of the city before and after" (http://www.semcity.net/cms/uploads/docs/STSMreport_final3.pdf) in a different direction. First, we gained some distance from the earthquake hazard relationship and focused on Modernist heritage protection. We kept the idea of having a guided tour and a game to identify landmarks (Fig. 7).

If we were to apply the concepts developed during the STSM in Faro to Magheru boulevard, we could chose a topic of interest for this book, namely not the built, but the unbuilt spaces, the free spaces, the gardens. A project of the Agronomy University of Bucharest proposed to link the free spaces around the Boulevard for an increased quality of urban public space, maybe in a way as we later propose for Pocket Parks. But we also propose something else. We propose to look at history. The first editor of this book, Cerasella Crăciun, had a project dedicated to "Lost gardens", gardens which had to make place to building. Most of them were located in this area. The archive research which ended with an exhibition could be used for digital routes. Instead of the buildings disappeared in the earthquake in Lisbon, the gardens which disappeared because of the construction project can be the landmarks.

Therefore, in the game, the landmarks will be identified with the simplified 3D models derived from their 2D sketches, as we can see, for example, in the Marcel Janco parcours developed by e-cart.ro (http://www.e-cart.ro/asociatia/ro/noutati/Traseu_urban_M.Iancu.pdf). Such an approach is aimed for all the architects identified in Sect. 2.2.

It is Marcel Janco for whom Augustin Ioan (2012) made an exercise of so-called morphogenesis, which is what we intended in the concept implemented in an exercise under the guidance of Daniel Libeskind at the University of Arts and Design in Karlsruhe called "Spaces to encounter" (Voigt et al. 2010). Outgoing from a 2D shape a 3D shape is created through morphogenesis. (Fig. 8).

We did such an exercise outgoing from a cube volume using the grey tones when converting a Mondrian painting to assess heights (Fig. 31b). In our exercise of utopia, we suggest also a building based on cube shapes (http://bostenaru.natkat.org/ucyw/archi/hhausz1.htm—a project of the first author at the institute of the second author at student times, a publication is under review for Acta Technica Napocensis), but the main finding related to this is the one already mentioned regarding the setting together of volumes in building urban models in "material" form or in Google Building Maker.

Looking at these morphogenesis exercises, we observe that the completed ones go out from the layout of the plan. However, in what we propose for Lisbon and also for the urban parcour, we should go out from the perspective drawing. In this, we shall follow volumes as set together in the building of urban-scale models (material or in Google Building Maker), while when going out of the plan, the interior space within such a volume of constraint shape (e.g. of a cube in our own) was done.

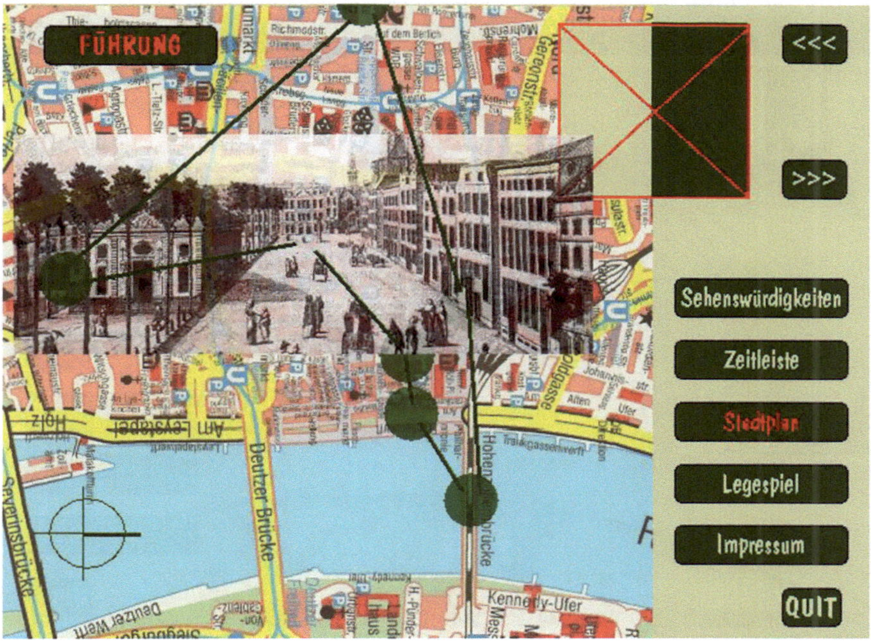

Fig. 7 Game for recognising current landmarks on their historical setting and guided tour for past landmarks on a contemporary map, example for the city of Cologne. Software application developed by Bostenaru Dan, 2000

It is possible to make thematic maps not only for the architects mentioned, but also for issues related to disasters, for example buildings resulting resilient from the cooperation between the architect and the engineer, a route included also in the study zone from this research and exemplified in Bostenaru Dan (2005b). Other thematic maps related to disasters may include the strategic elements for disaster management, aiming at the morphology of the city, as explained by Florescu (2006).

Coming back to what we mentioned in doing the works, 3D printing follows the way it used to be drawn in architecture, with solid walls, which were filled with black ink colour at the time of ink drawings, to emphasise the space, in an approach related to the above-explained morphogenesis. The building of models at detail scale followed, however, what is essential for "materiality": the construction, something involved by going over from chapter architecture to the built object on the site. The walls had to bear the ceilings. Such logic is also followed when developing toys such as a 3D puzzle.

This developed concept is compatible with a contemporary approach in urban planning which is called "strategic planning". As opposed to traditional planning where all buildings and all city zones are treated and detailed equally in the master plan, in strategic planning, some of these deserve more attention, taking the shape of pilot projects. In our approach to the case study of a Modernist neighbourhood in northern Bucharest, we applied strategic planning: for directing the

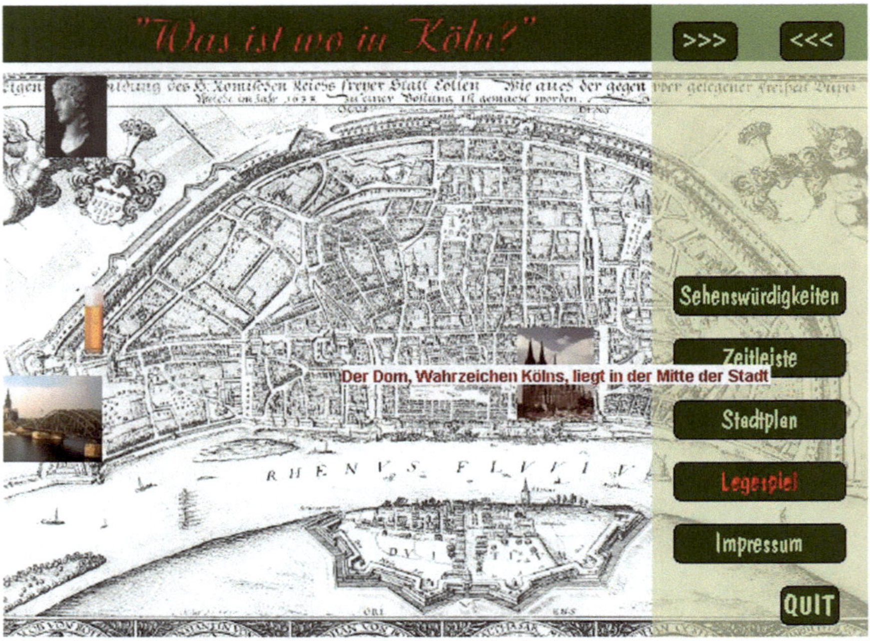

Fig. 7 (continued)

development, we developed a vision, for which there are action plans at different levels of planning from the urban level to that of the building element over the level of building, instead of just different urban planning levels (Bostenaru Dan 2013). But the key elements in our approach to strategic planning are not the social aspects of voluntary agreements in the democratisation of planning which is participatism but the approach to the problem, with its many facets. For this, the SWOT analysis has been approached in an innovative way not defined here, following the way Lynch (1960) read the "image of the city". This project in itself can be a demonstration or pilot project. The next level on which we see innovation is the one in regulation and financial undertaking. The regulation we propose is drawn on the basis of the innovative analysis. And one of the actions to achieve the objectives of the strategy is the establishment of an environmental budget. The final step to achieve innovation is the education, which has been designed as a feedback mechanism. In Bostenaru Dan (2004), we saw this Kevin Lynch approach.

We explained in an article regarding multi-criteria decision how these action plan levels relate to each other (Bostenaru Dan 2004). In that figure, we showed that from the urban zone level, there shall be a zoom to the building level and from the building level a zoom to the building element level. We followed that principle of zooming when presenting our work in HTML with the help of image maps (http://bostenaru.natkat.org/ucyw/archi/bauf1.htm or http://bostenaru.natkat.org/ucyw/archi/hhausz1.htm) created using traditional drawing and model creation.

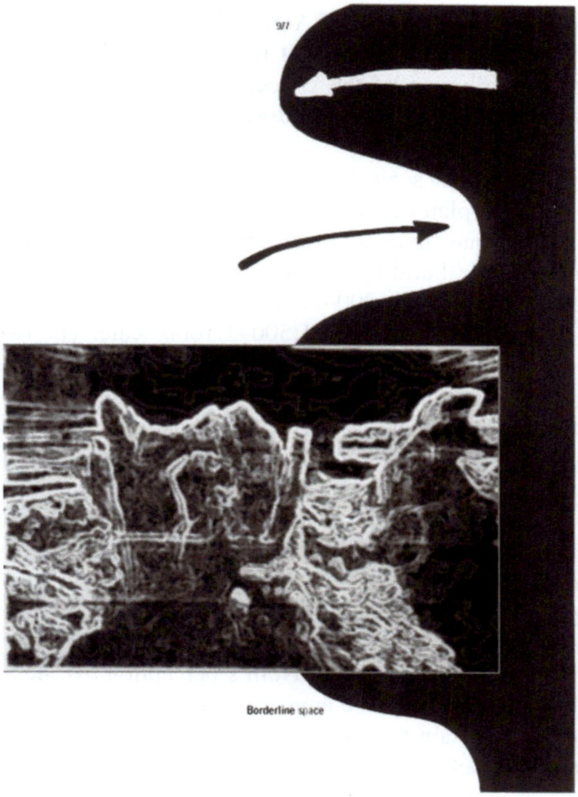

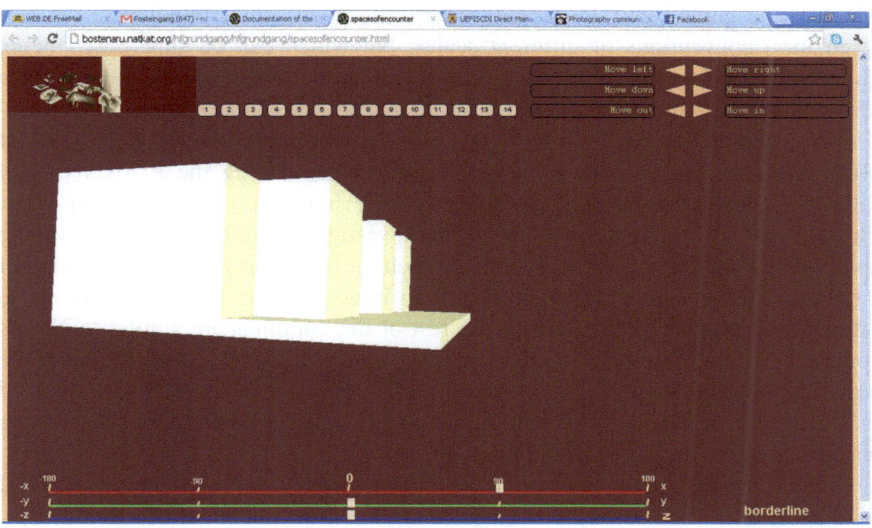

Fig. 8 Modelling approach for the landmarks. Exercise for a borderline space, within the "Spaces of encounter" seminar conducted by Daniel Libeskind in 2001 at the University of Art and Design in Karlsruhe. 2D collage and 3D model in real-time software rendering by Maria Bostenaru

However, with today's techniques in CAD drawing and 3D printing, these issues of scale are not so important anymore. It is the contribution of digital methods. This is the main contribution of the concept developed during this Short Visit Grant at KIT compared to the concept in Bostenaru Dan (2004): the elimination of scale issues.

In our approach against scale issues, we are supported by such an ancient concept as the baroque plan of Rome by Nolli (http://nolli.uoregon.edu/), coming thus back to Baroque times as in the initial investigation about Lisbon. In the plan of Nolli, public buildings are presented with the detail of plans, just like a model of a building in 1:200–1:50 scale, while common buildings are a black mass, just like urban models in 1:500–1:1000 scale. The two scales come together, as they can come in GIS layers today, when we can zoom with computer technology.

We analysed in several papers of us (the mentioned Bostenaru Dan 2004 paper and a paper on the 3D representation of the street Bostenaru Dan 2011b) how the elements defining for the urban image can be highlighted outgoing from photography. It is also what Building Maker uses when simplifying the photography to simple volumes and what we intend for going 2D to 3D in simple shape for the 3D model of representative buildings. What is additionally shown is also the language of the space like in the Libeskind exercise—a language at the level of the urban space, not of the interior space, based on the same perception language. The correlation between the two leads us to Nolli's perception. At the same time, this concept of the combination could be applied for urban assemblies as those identified as monuments, or certain features of the volume can lead to recognising landmarks or zones as in Lynch's (1960) description.

Therefore, we propose for the second part of the scenographic application, which is the guided tour, to have different GIS layers, corresponding to the representation conventions of different scales, but which can be superposed being at the same scale. The buildings setting together the route (being those of one of the chosen architects or the buildings damaged by the earthquake, or the successful collaboration between architect and engineer) are detailed at the level of plans. We investigated how to represent such plans during the PIANO project (http://bostenaru.natkat.org/) and which is their importance and their typology. Through this, we make an addition to that investigation, done mainly at building level, where we differentiated between attaching the information to the space of a room or to the building element, again an issue of scale, but also of timing in estimating the future project costs. The rest of the buildings are similar to Nolli's common buildings. It is a difference to the approach by Sitte (1889) where the buildings were highlighted in different shades of grey and black, but none detailed at the level of the plan as for Nolli. Sitte's (1889) approach is closer to what can be performed with the help of traditional GIS.

We observe that in the guided tour apart of the dots from the figures in part 2.2 or in Bostenaru Dan (2005b) which can be replaced by scale plans of the buildings, also the photograph can be included, to be blended when scrolling above, the advantage of an interactive movie.

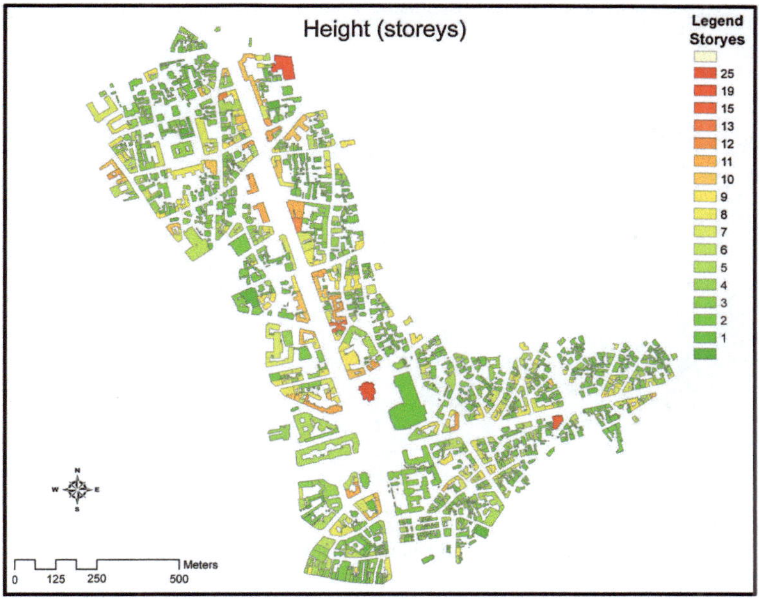

Fig. 9 Height analysis of the Modernist boulevard in Bucharest using GIS

4 Further Development Possibilities: A Discussion

A first further development possibility regards the combination of GIS and Google Earth analysis with regard to what has been analysed in the Short-Term Scientific Mission we pointed to. Some of these combinations allow the implementation of our concept. Figure 9 shows the Modernist boulevard in Bucharest which was investigated, displaying the GIS analysis considering heights. In a publication of us, we did an analysis to which extent GIS models are suitable for the depiction of 3D city models (Bostenaru Dan et al. 2013). However, what we learned in frame of the NeDiMAH network was the dialogue between GIS and Google Earth. In Google Earth, shapes can be modelled in detail using SketchUp, or textured with Building Maker, different from the 3D extrusion done in GIS (with the software developed by Armaş et al. 2010 or with the ESRI software CityEngine). This would correspond to the detailed views we proposed in our concept, since with SketchUp, it is possible to also model the interior of the buildings, that is, the plan. For Bucharest, this is especially important since there are no 3D models of the city to our knowledge, and only a few buildings, landmarks, have been modelled by enthusiasts using Google Earth. Figure 10 shows this superposition.

We developed a concept to represent in 3D the Modernist boulevard, different of CityGML but considering different levels of detail. It is an innovative approach, but more research is needed also to implement it. In a further research of us, we will focus on datascapes created by the dialogue between GIS and Google Earth,

Fig. 10 The Modernist boulevard in Bucharest 3D extrusion in GIS (city engine) seen with a digital browser, Google Earth, including the highlighted models from this

which is directly related to this research. Datascapes is an innovative concept introduced by MVRDV (1999), a successful architecture office aiming to map data connected to urban life. The title of the book, "Metacity/Datatown", supports digital network analysis where the connection between the items is made through their metadata. Trying to apply 3D extrusion to the data contained in our GIS database, a first observation is that if GIS is able to map also qualitative analysis stored as strings, the datascape can be realised only for numerical values. As such, we made a datascape for function where only the residential buildings are mapped, based on their area (Fig. 11). A software developed by Nadia Amoroso (http://www.datappeal.com/) allows to superposed more of these datascapes than CityEngine extrusion can do, making a 3D city model for any of the characteristics linked to the building ID. Amoroso (2010) presents the background for this. If the buildings lose their plan shape and are reduced to centroids (Fig. 12), then one can map more characteristics on the same map. This is the transposition of urban analysis as it has been done on paper (Fig. 13) and as it is not allowed by GIS which does not have this language of highlights. In the public database of DataAppeal (http://www.dataappeal.com/data-gallery-2/), there are a couple of superpositions of this kind related to landscape planning for example "diabetes rates and tree coverage", or "vacant parcels and wooded lots", "Berlin trees, parks and open space", "parks and open space trees", "traffic collision and pedestrian flow" related to our topic but also for single interpretation such as "green roofs", "tree coverage", "age of trees", "the damaged urban forest", "green open space", "Great Lakes parks", "Ottawa earthquake", damaged trees pre—and post—Sandy strom.

Related datascapes, but for earthquake vulnerability, can be done in Google Earth using the software SELENA (Molina et al. 2010), and their employment for the Modernist boulevard is an issue for further research.

Fig. 11 Datascape of the function (residential area) of the Modernist boulevard

The functionality of the street network is especially important in this area because it includes a major number of buildings in risk category I regarding earthquake vulnerability. The GIS model we discussed was part of the EQSIM model developed in frame of the SFB 461 which also analysed, agent-based modelling on how search and rescue teams and other emergency management-related tasks can be carried out in post-disaster intervention related to the street network. EQSIM also did a conversion to a 3D model in Google Earth (Zikas and Gehbauer 2007). An alternative to Space Syntax analysis of the flows in the street network is agent-based modelling of the flows on the street. This kind of analysis has been done by Fiedrich (2004), proposing a high-level architecture model for a GIS inventory of the streets in the area. Frank Fiedrich (2004) first investigated the state of the art regarding the employment in post-disaster management of operation research, expert and decision support systems and simulation systems. He employs the belief–desire–intention model and the corresponding high-level architecture in the software environment. The HLA encompasses a simulation component and a decision component. The decision is based on the action of agents. There are several agents active in post-disaster intervention such as FireAgent, SARAgent (search and rescue of victims), AmbulanceAgent, ReconAgent (reconnaissance), InfraAgent (for the street infrastructure). The implementation is strongly dependent on the later, since the number of victims from the simulator depends on the damage to buildings (recon), and the way ambulance can move depends on how

Fig. 12 Datascape centroids for an analysis of more criteria simultaneously as in urban planning site analysis

much the streets are filled with debris. The agents communicate with each other. They are fed with plans like in an expert system. This means that each agent tries to carry out a prefilled plan, and only if the plan is completely carried out, it is successful; otherwise, the next plan is tried (action layer, coordination layer, planning layer). We note that the fire was considered for the agent architecture, since it has a linear spread like in games (SimCity), instead of the earthquake which damages buildings randomly. Agent-based design can also represent the substance of the city, which is not detailed at Sitte (1889) and Nolli and as Nolli‘s map from 1748 as well as Patrik Schumacher in his lectures on parametric semiology shows.

An original approach which deserves further investigation is the interaction between GIS and hard copy models, such as those mentioned by us when analysing how 3D can be printed. The digital which served for the print and original digital can interact. A company in Portugal developed a projection system of GIS that highlights certain areas on terrain models (http://www.gison3dmap.com/) (Fig. 14), which can be also applied for urban areas, and is thus useful for our system of highlighting the landmarks in traces. The method of interaction between projection and a hard copy 3D model is also used in the Lisboa story centre for highlighting what we tried to do in this work: the parcours and the trace of different events—and being used for historical events including the Lisbon earthquake of 1755 connects to our earlier research.

Fig. 13 Urban analysis in the same way as datascape. Analysis for the port of Frankfurt on the Maine. Diploma work of the first author at the institute where the second author was academic councillor at the time

Also, talking of urban routes, we did an analysis of the street network. For our research, we considered the analysis proposed by Space Syntax, an internationally recognised approach, and the respective software to highlight the

Fig. 14 GIS on 3D map for the cultural landscape in Sintra—exhibition in the park. Photo by M. Bostenaru in 2013

interdependencies in the street graph in the area (Fig. 15). We used the software AJAX (Batty 2005). The street becomes an element of a complex network in the context of Space Syntax analysis. Historical network research is a young and dynamic field to face the challenges by social network analysis in historical time and has been set up only recently. However, Space Syntax has been developed by Hillier (1999) around 1980 with a similar aim: to simulate the social effects of design. Bafna (2003) gives a more recent overview of it since there are biannual symposia to its employment. The connectivity of spaces can be seen as choices a pedestrian makes when walking in the city, and it can be thus the choice to be followed when doing an urban route. Mihăilă (2014) analysed such architecture routes developed for Bucharest, mainly as a part of cultural tourism. We can say that the city is seen as a museum. The exploration of the urban neighbourhoods can be done searching for these museum elements, but can be done also looking for the atmosphere, Mihăilă (2014) said, as we have done in an excursion at the Canadian Centre for Architecture to explore Montreal neighbourhoods and in which, for example, we discovered the dialogue between plants and architecture as shown in the chapter by Bostenaru in this volume. Dursun (2007) analysed Space Syntax exactly for this, for museums, but not open street ones, instead for architectural objects.

One can analyse Space Syntax at the level of the floor plan, as we have did developing an ontology in our PIANO project (http://cordis.europa.eu/projects/rcn/89804_en.html), in the zoning of the building, or, more frequently, at the level of the urban space. Vialard and Bafna (2009) presented how Space

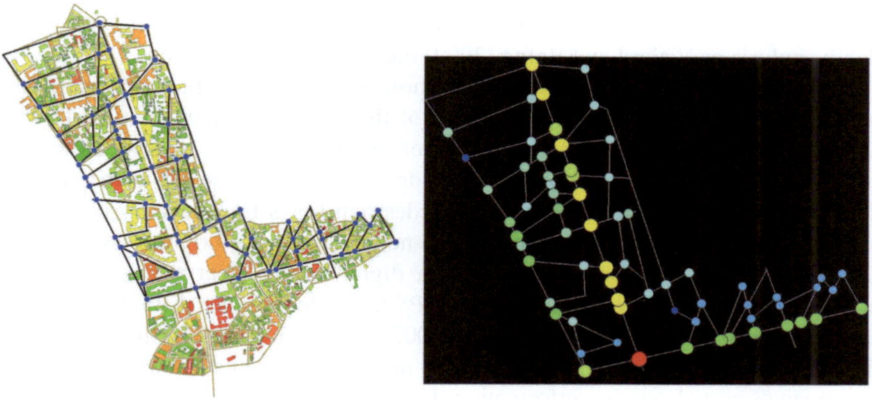

Fig. 15 Junction analysis for the Modernist boulevard, 2012. Computed with Space Syntax software AJAX

Syntax applies to interwar housing like the one investigated in the PIANO project and featured in majority by the Modernist boulevard we analyse in this paper. In the PIANO project, we analysed zoning and developed an ontology for the relationship between spaces. Exploitable foreground includes the following:

- The idea of ontology for zoning, which can be used for the development of object-oriented software dedicated at project management in construction industry, other than based on spaces or building elements as it is now. Ontologies are a way of organising data types for object-oriented programming. Right now, the software for facility management is based on project management, so this would be a new approach. Additional work includes the definition of taxonomies and the creation of a database to try out the concept. The two exploitable foregrounds are connected by the sense of ontologies/semantics, but differ in the field of application.
- Relationship between function and sign (semantics, as seen by Umberto Eco, philosophy) applied to the zonification of this housing, to be connected with the ontology. Eco (1997) used ontology and semantics in its philosophic sense. In order to make it an exploitable foreground, the definitions have to be converted to those used in computer science, especially semantic Web. This would be a further work compared to the project and might be done during the work done for photography with similar network research concepts (Bostenaru Dan 2011c).

In the concept we developed in this paper, we link the floor plan to the city plan as Nolli did for public spaces, as in Dursun (2007). The analysis Vialard and Bafna (2009) can be thus applied for spatially linked buildings, since at the same location, to show evolution in the intervention on the street. The urban and the building scale can be thus integrated in the same Space Syntax analysis using a Nolli type display for the Modernist buildings as developed in our concept superposing the depth analysis as Vialard and Bafna (2009) did on this representation.

Extrapolating from the floor plan, the choices in the urban space are made also on the basis of zoning, which is division of the urban settlement into parts with a unique function. Space Syntax analysis of the flow in urban streets can lead to choosing the best route to connect the landmarks identified so far in the chapter. There are different ways of monitoring pedestrian flows for Space Syntax analysis. We would like to propose, however, a new way for our urban routes project. COMOB (http://www.comob.org.uk/) is a digital arts project that explores the potential for collaborative mapping with GPS technology. It is a different approach to public participation GIS (Craig et al. 2002), a direction which gained attention from the idea of empowering marginalised populations such as aborigines. PPGIS is a mean of crowdsourcing information. However, Space Syntax can be connected with GIS. Jiang and Claramunt (2002) critique the axial view Space Syntax proposed and instead propose a network view with visibility lines. In a workshop (Amoroso et al. 2013), we had the occasion to map using the COMOB technique (Fig. 16). To each item perceived as a landmark by the mapping participants, an image and a text can be associated via iPhone and then mapped by the central application to an urban route. It is an alternative way to a participative method which leads to PhotoVoice, where citizens are seen as "sensors" mapping collaboratively the landscape on the way to volunteered geographic information. Just that in our concept the items on which information shall be mapped are given, there are only the traces between them, which might be different. Such routes we propose to then analyse via Space Syntax in order to gain new routes in the built substance which can be emphasised by new ways in the landscape (different pavements for example) linking Pocket Parks (Fig. 17). In another project started by the first author at the department of the second author and with an urban focus under the guidance of Henri Bava, the mineral element and the vegetal element form patches which interfere creating the trace itself set over the urban texture and leading from the train station to the centre. It is a way of creating nervous interconnections in public space and thus a mind map addressing memory. In this case, not the urban texture, but the texture of materials has been symbolically approached in the public space of pedestrian streets. The given urban texture does not display coherence, and a network trace is necessary in order to make the textures cooperate.

Bostenaru Dan (2010a) presented a way of designing the paradise garden as a labyrinth. To describe this, Bostenaru Dan (2010a) writes:

> As a mesh in the fabric of life, the moment is not something isolated, but contours the door to what is located behind the life. The life, consisting of independent meshes, becomes a series of gates that lead to parallel passages which also led up to others. A particular moment is an unexpected gate that leads to the mystical for us.
>
> The lace fabric of life, consisting of independent meshes, disappears at the moment when we find the door to what shone before behind life. We direct our steps to what lay behind the world out, but without having to leave this world. We step as on a smooth, clear glass, and from down on the eternal abyss follows us.
>
> The inner world is what lies behind life, the virtual. This can only be achieved by the mean of feelings. They each form unexpected doors.

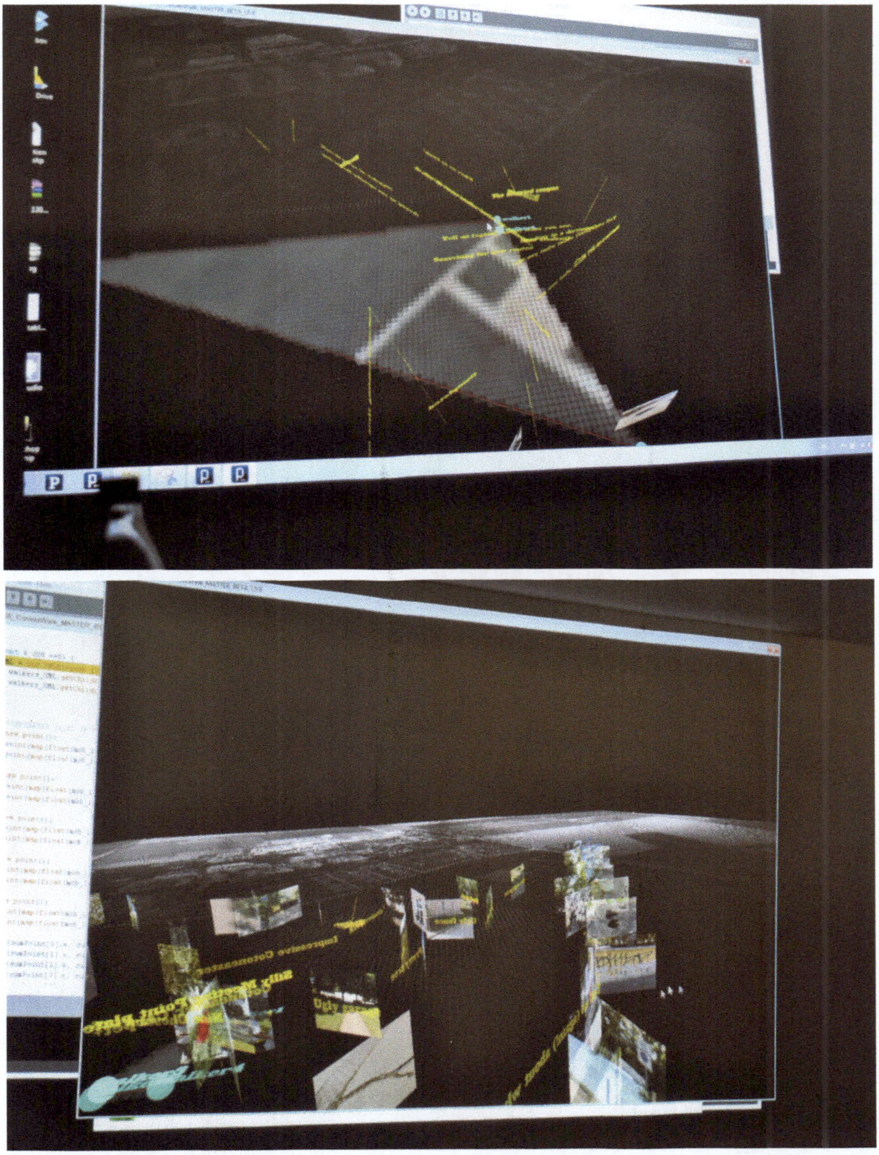

Fig. 16 Patchscape of COMOB workshop at the digital landscape architecture conference 2013 in Bernburg

Translated to our concept, the labyrinth is the network of traces followed in the city, where at each node, choices wait and there is no Ariadne thread file. The project in Fig. 17 shows such a garden which attempts to transpose the paradise garden in everyday life. Feelings are suggested through the morphogenetic modelling

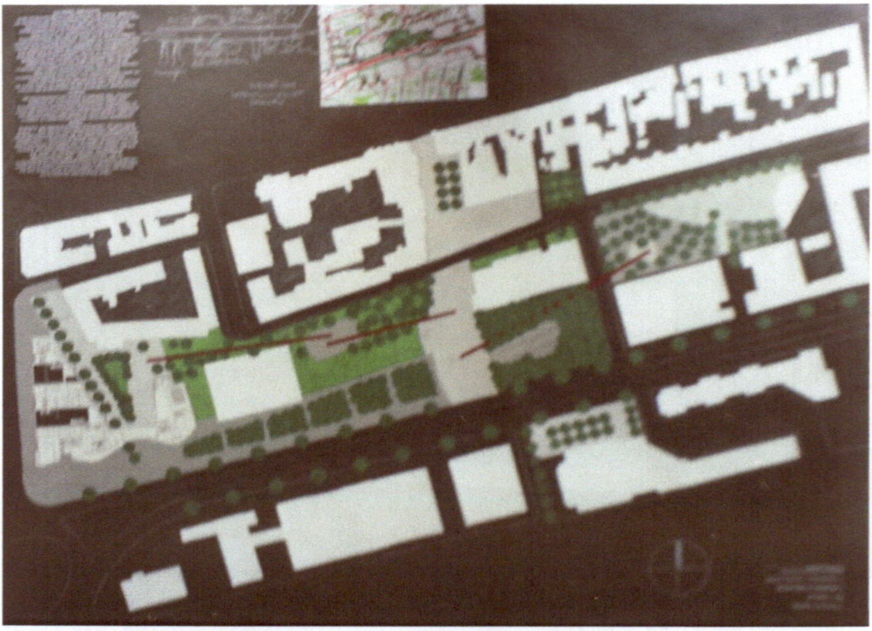

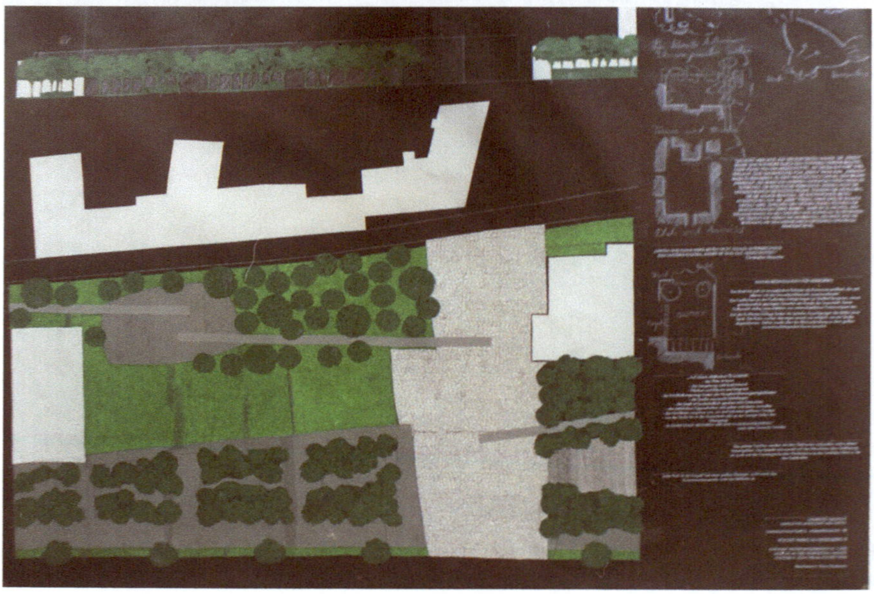

Fig. 17 Mineral ways in order to build a network of pocket parks (of vegetal public spaces) visualised by means of collage. The way from the train station to the centre in Heidelberg, Germany. Project by Bostenaru Dan, 1998. The architecture project was done at the institute where the second author was back then, and please note that the code proposed in this paper is being used: the architecture project is detailed as floor plan as we propose and as Nolli did for public buildings, while the the other buildings are represented as the common ones. This way the environmental centre has the same public role as the designed landscape

of buildings in the way presented in the Libeskind exercise, and thus, they are doors to a rediscovered space of the history. In fact, in the exercise, these boxes were indeed suggesting feelings through the items contained in the LandArt model from which they derived (Bostenaru Dan 2010a), which was in fact a door to a chosen space from the exercise, the "rediscovered space". Gardens can form green patches also in the discussion Henry and Frascaria-Lacoste (2012) make on green reconciliation ecology.

GPS data are metadata and can be included in metafiles, both of already available photographs as those gathered by the first author and presented in the first paragraph of this paper partially, but also of SketchUp models as envisaged by the morphogenesis in the paper, for example for inclusion in Google Earth by means of superposing GIS and Google Earth (for example, with CityEngine).

Cosmin Chirvasie developed a cybernetic analysis of cities as a system which led to the 3D modelling through fractals of the historic evolution of organically born cities. He modelled how the highest points in Middle Age Vienna, for example, were dictated by the street network (http://myriammahiques.blogspot.com/2010/02/introducing-arch-cosmins-work-on-urban.html). Street and building are interconnected in a complex graph. In his research, the street dictated the position of the landmark and not the landmark how the route is born. Barthélemy and Flammini (2008) connected Space Syntax to the way Chirvasie conducted research. Ratti (2004) notes in this context that Space Syntax is rather applicable for organically grown forms and not for grid structures, as is the fractal development of Chirvasie. It depends on how the space is striated and preplanned city (Bostenaru Dan 2010b) or reconstructed ones like the Baixa new centre in Lisbon would less suitable for mapping pedestrian choices.

Returning to the idea of sign and semiotics which we developed in connection with Eco (1997) approach, a recent Marie Curie project (EastWest) is dealing with "A study on Space and Language: Relationships between Ideograms, the Roman Alphabet, and Urban Space in the East and the West" (Salvator-John Liotta and Philippe Bonnin, http://cordis.europa.eu/projects/rcn/108684_en.html http://prezi.com/efevuhwbegsm/untitled-prezi/). In this project, the city is not read, but written in connection with the alphabet, and it is a historic view. Preliminary research has been conducted by the group of Philippe Bonnin for cities in France and China and Japan in the project "Modèles Numériques de Morphogénèse Viaire" (Numerical Models of Street Morphogenesis, http://www.morphocity.fr/), using again the street network as a means for historic analysis, as Chirvasie. Also today, cities are written when we read them, since new routes are born, as those designed in this project.

For this first study, similar items at the same location have been connected by means of spatial graphs. For further research, we propose to look at connections between the networks proposed as an abstract network based on semantics. A beginning of looking to semantics has been done with mentioning the sign of Eco (1997) and the research in the EastWest project. Such links can provide for a subject-predicate-object relationship to be used in other network software (for example, ORA) in order to connect not only by site, but also by typology of the investigated spaces in the city or morphologies of the buildings (as envisaged by the morphogenesis proposed in the concept).

5 Conclusions and Future Work

This paper presents a contribution to historical network research from the point of view of architecture. A semantic network can be built by means of the concept proposed here between buildings of the Modernism, investigated many times separately in the history of architecture, separated from the site and separated from similar developments in other countries. Also, the floor plan tends to be neglected, what will be overcome by the research proposed here, which integrates the interior of the building in the spatial analysis of the whole city, in a Nolli type representation, in highlighting the models in connection between GIS and Google Earth and in Space Syntax analysis of Modernist buildings. This will make possible comparative research at architectural level and at urban level, between the elements of the network, between the buildings and between the network itself as urban landscape. Cross-location analysis can be performed (actor x actor) for mapping the dispersion of certain architectural approaches across Europe. Alexander et al. (1977) created a network of patterns, but for the purpose of mapping, one can compare to the citation network of some researcher.

Also, the creation of new paths to connect the "museum" objects which the elements of the urban trace build will need landscape architecture interventions. The street network becomes a labyrinth in which choices are made depending on the landmarks of each particular mind map according to Lynch (1960) analysis. The text we enter via participative investigation can form a tag cloud/word cloud (weighted list in visual design). The Lynch analysis is particularly suitable because actually such an analysis type in a work of us led to an approach similar to that of the so-called protected zones in Bucharest before these were defined (Bostenaru Dan 2013).

Such an approach can be applied in investigating the perception of landscape in video recording or sequential recording, with application to landscape design teaching, which is the subject of current doctoral work at the first author's department (work of Irina Pață, personal communication).

The paper also presents an innovative approach of dealing with GIS layers by including more scales, adequate for strategic planning in urbanism, and opposed to traditional rule studies.

We introduced to further studies to connect to the history of the city as described by the evolution of the street network. This not only dates the buildings by when the parcel was created, but also, as reviewed in Chirvasie's research, determines the best place for the landmark. It is to be analysed like in project management if the critical network path has a role in this and thus establishes a new direction.

The applicability of the concept has been tested so far for urban structures organised in blocks, such as those promoted by Sitte (1889), and evidently for

European urban structure, as it is one of the Modernist boulevards. Applicability for other Modernist developments as the Siedlungen in Western Europe, which follow Le Corbusier's ideas from Plan Voisin and Ville Radieuse, needs further investigation.

Acknowledgments This paper is based on work developed within the NeDiMAH (Network for Digital Methods in Arts and Humanities) Short Visit Grant 5454 of the main author hosted by the second author with the title "Architectural heritage protection of the central area of Bucharest—mapping ways of visualisation in GIS and archives (so-called "registries" or taxonomy/ontology entries)". The work was presented at the conference "The future of historical network research" in Hamburg, Germany, also funded by NeDiMAH and includes updates to the report based on the comments received. Initial routes have been done during field trips in frame of the Marie Curie Fellowship funded projects CA'REDIVIVUS and PIANO in Italy and Romania. Some ideas on taxonomy and agent-based design have been analysed in the frame of the Short-Term Scientific Mission COST-STSM-ECOST-STSM-IS1104-010413-024718 at the ISCTE Lisbon, Portugal.

References

Alexander Ch, Ishikawa S, Silverstein M (1977) A pattern language, towns, buildings, construction. Oxford University Press, New York

Amoroso N (2010) The exposed city: mapping the urban invisibles. Routledge, London

Amoroso N, Hudson-Smith A, Phillips M, Speed C, Willis K (2013) Concept for workshop: patch-scape; pads, pods, phones and spatial data. In: Buhmann E, Pietsch M (eds) Digital landscape architecture. Wichmann, Berlin, pp 330–341

Armaş I, Dardala M, Reveiu A, Furtuna F (2010) Spatial modeling of urban environmental vulnerability to seismic risk. Case study: the historical center of the Bucharest municipality—Romania. In: Proceedings of the WSEAS/CIEO international conference on urban rehabilitation and sustainability, Faro, 2010

Bafna S (2003) Space syntax, a brief introduction to its logic and analytic techniques. Environ Behav 35(1):17–29

Barthélemy M, Flammini A (2008) Modeling urban street patterns. Phys Rev Lett 100:138702

Batty M (2005) The AJAX project: new theory and new software for space syntax. In: Proceedings of the 5th international space syntax symposium, Delft, The Netherlands, 13–17 June 2005

Bostenaru Dan M (2013) Management of innovation: innovation policy in the urban development. Urbanism Arhitectura Constructii 4(2):3–18

Bostenaru Dan M (2012) Arhitectura interbelică cu structură din beton armat expusă la hazard seismic în context european. Intervenţii în spaţiul romanesc şi Italian, doctoral thesis, "Ion Mincu" University of Architecture and Urbanism, Bucharest

Bostenaru Dan M (2011a) Seven years dedicated to the conservation of the Modern Movement Heritage, e-conservation magazine 18, pp 19–26, http://www.e-conservationline.com/content/view/977

Bostenaru Dan M (2011b) Street network issues for disaster management. In: Proceedings of the 11th international conference "Reliability and statistics in transportation and communication" (RelStat'11), Riga, Latvia, 19–22 Oct 2011, pp 443–452

Bostenaru Dan M (2011c) The use of ontology for digital conservation of architecture works after catastrophes. J Appl Eng Sci 14(2):11–18

Bostenaru Dan M (2010a) Spaţiul verde redescoperit—Der wiederentdeckte Grünraum. Cuvillier Press, Göttingen

Bostenaru Dan M (2010b) The flat and the knurled space in the reconstruction of cities. In: Panangopoulos T, Noronha T, Beltrao J (eds) 3rd WSEAS international conference on advances in urban rehabilitation and sustainability, Faro, Portugal, pp 80–85

Bostenaru Dan M (2005a) (Un)Sincerity in the exterior expression: spatial structures in the avantgarde architecture. In: Mircea Mihăilescu and Călin Mircea (eds) Proceedings of the international symposium on shell and spatial structures, vol II. Editura Mediamira, Cluj Napoca, pp 753–760

Bostenaru Dan M (2005b) Multidisciplinary co-operation in building design according to urbanistic zoning and seismic microzonation. Nat Hazards Earth Syst Sci 5:397–411

Bostenaru Dan M (2004) Multi-criteria decision model for retrofitting existing buildings. Nat Hazards Earth Syst Sci 4:485–499

Bostenaru Dan M, Panagopoulos T, Gociman CO, Armaş I, Dill A, Chiriloae A, Florescu T (2013) The impact of hazards on the urban tissue—3D representation and digital databases. Adv Geosci 35:45–53

Condaratos S, Wang W (1999) 20th century architecture: Greece. Prestel, München

Craig WJ, Harris TM, Weiner D (eds) (2002) Community participation and geographic information systems. Taylor and Francis, London

Dursun P (2007) Space syntax in architectural design, 056. In: Proceedings, 6th international space syntax symposium, İstanbul

Eco U (1997) Function and sign: the semiotics of architecture. In: Leach N (ed) Rethinking architecture: A reader in cultural theory. Routledge, London, pp 182–202

Fiedrich F (2004) Ein High-Level-Architecture-basiertes Multiagentensystem zur Ressourcenoptimierung nach Starkbeben, PhD thesis, University of Karlsruhe, Germany

Florescu T (2006) Formă şi Trans-Formare Urbană. Editura Universitară "Ion Mincu", Bucharest

Gallagher H, Farmer B, Mendoza C, Lee C, Dickson H, and Greene M (2013) GEM building taxonomy v2.0: evaluation and testing report. GEM building taxonomy global component, available from http://www.nexus.globalquakemodel.org/gem-building-taxonomy/posts/report-on-evaluation-and-testingof-the-gem-building-taxonomy-released/

Henry A, Frascaria-Lacoste N (2012) The green roof dilemma—Discussion of Francis and Lorimer (2011). J Environ Manage 104:91–92

Hillier B (1999) Space is the machine: a configurational theory of architecture. Cambridge University Press, Cambridge

Ioan A (2012) Marcel Iancu şi alfabetul său formal: un exerciţiu didactic în derulare (I), Arhitectura 3/2012

Jiang B, Claramunt C (2002) Integration of space syntax into GIS: new perspectives for urban morphology. Trans GIS 6:295–309

Le Corbusier (1971) Chartre d'Athenes, Seuil, Paris (first publication 1943)

Lynch K (1960) The image of the city. MIT Press, Cambridge

Machedon L, Scoffham E (1999) Romanian modernism: the architecture of Bucharest, 1920–1940. MIT Press, Cambridge

Marat-Mendes T, Sampayo M (2010) Étienne de Groer: The Scales of Urban Intervention in the Lisbon Territory. In: Proceedings of the 1st European architectural historians network conference, Guimaraes, Portugal, pp 32–39

Mihăilă M (2014) About possible architectural routes and lectures. Bucharest—a short study, Urbanism, Arhitectură, Construcţii. Online first

Molina S, Lang DH, Lindholm CD (2010). SELENA—an open-source tool for seismic risk and loss assessment using a logic tree computation procedure. Comput Geosci 36(3): 257–269, http://dx.doi.org/10.1016/j.cageo.2009.07.006

MVRDV (1999) Metacity/Datatown. 010 Publishers, Rotterdam

Penelis G, Penelis GR, Paschalidis KP (2003) The structural upgrading of the army pension fund (A.P.F.) building in Athens, FIB2003, Athens, 2003

Ratti C (2004) Urban texture and space syntax: some inconsistencies. Environ Plan B. Plan Des 31:487–499

Sharp D, Cooke C (2000) The modern movement in architecture—selections from the DOCOMOMO registers. 010 Publishers, Rotterdam

Sitte C (1889) Der Städtebau nach seinen künstlerischen Grundsätzen, New edn. Birkhäuser, Basel 2002

Sonne W (2009) Dwelling in the metropolis: Reformed urban blocks 1890–1940 as a model for the sustainable compact city. Prog Plan 72(2):53–149

Vialard A, Bafna S (2009) Syntax of change in the mid-twentieth century American house. In: Koch D, Marcus L, Steen J (eds) Proceedings of the 7th international space syntax symposium, KTH, Stockholm, Ref 130, 130:1

Voigt A, Aharoni S, Birken J, Krell A, Zappe J (eds) (2010) The Daniel Libeskind research studio. Staatliche Hochschule für Gestaltung Karlsruhe, Karlsruhe

Zikas T, Gehbauer F (2007) Decision process and optimisation rules for seismic retrofit programs. In: SFB 461 (ed) Proceedings of the international symposium on strong Vrancea earthquakes and risk mitigation. Bucharest, Romania, pp 472–484

Landscape Rehabilitation Through Policies and Urban Interventions for Reshaping Mobility

Mihaela Hermina Negulescu

Abstract Improving the quality of life in urban areas, including its landscape dimension, became a common objective of development strategies for all urban/metropolitan areas, reflected in their mobility policies. The majority of big cities undertake more and more extensive urban operations for reshaping mobility and its infrastructure, which often generate the most complex dynamics of urban landscape renewal, from the establishment of intermodal hubs or organizing network of trams, which are the central axis of major landscaping redesign (Strasbourg, Paris, Freibourg, Los Angeles etc.), to the partial or total covering or burial of large roads (Paris, Boston etc.); they bring unprecedented opportunities for landscaping large urban spaces and areas. The renaissance of "urban boulevard" concept shows the willingness to reconcile and reharmonize all different hypostasis of the street: from that of infrastructure (corridor of movement) to those of urban landscape and community space, generating functional and spatial restructuring of important roads, with marked landscape valences.

Keywords Landscape rehabilitation • Urban interventions • Mobility • Transport

1 Introduction

Urban and environmental crisis, which since 1970s was acknowledged, reviewed, and affirmed as argument for changing the paradigm of human development, has been characterized, inter alia, by an acute deterioration of environmental quality of urban life, as a consequence of the unexpected, explosive, and unrestrained

M. H. Negulescu (✉)
Faculty of Urban Planning, Department of Urban and Spatial Planning, "Ion Mincu" University of Architecture and Urbanism, Academiei 18-20, 010014 Bucharest, Romania
e-mail: mihaela_negulescu@yahoo.com; ozifrumoasa@gmail.com

C. Crăciun and M. Bostenaru Dan (eds.), *Planning and Designing Sustainable and Resilient Landscapes*, Springer Geography, DOI: 10.1007/978-94-017-8536-5_13,

evolution of automobility.[1] The policy for "city adaptation to the car," in the sense of granting an unconditional priority of the latter, produced a suite of dysfunctional effects of various natures, from which the most directly noticeable is the deterioration of the urban environment in terms of image and esthetics—as cityscape—but also in terms of ambiance, of sensory perceptual and psychological impact that has people interaction with and in urban areas, abused, de-structured, and devoured by a transportation infrastructure (roads, highways, bridges, and railways) foreign to the scale of the traversed urban fabrics, by enormous, noisy, polluting and threatening motorized flows, by a tide of vehicles parked everywhere (see Fig. 1).

Therefore, in recent times, raising the quality of the living environment, including its landscape side, has become a common objective of all the strategies for large cities development, directly reflected in their mobility policies, as well as in more extensive urban interventions for mobility reorganization, which creates complex opportunities for cityscape rehabilitation.

2 Definition of Terms

Mobility is the *system of movement* in the territory, result of *travel behaviors* that fulfill *needs of travel* by *options* within an *urban travel offer* (urban form, transport, economic, and sociocultural patterns) (Negulescu 2011a).

Landscape means an area, as perceived by people, whose character is the result of the action and interaction of natural and/or human factors.[2]

3 Landscaping Requirements Reflected in Mobility Policies

One of the general objectives for remodeling mobility in the sustainable development logic is *the* "urbanization" *of the movement, through a non-confrontational insertion of its flows system into the urban territories, fabrics and landscapes* (see Fig. 2).

The requirements for protection of architectural values, of urban landscape, as community heritage with cultural, social, and economic value, generated, as did good accessibility requirements, the simplest programmatic syntagm of reshaping mobility: "the right move in the right place." The reorganization of accessibility (its various types), according to the traversed urban territories, fabrics, and spaces, marks the exit from automobility sovereign period and the transition phase toward a new aspiration, for re-"humanization" of cities, by finding a (more) balanced

---

[1] The use of automobiles as the major means of transportation.

[2] European landscape convention, Council of Europe, Florence, 2000.

Fig. 1 Impacts on the urban space of the policies for unrestricted "adaptation of the city to the car." **a** Big, even huges transportation infrastructures foreign to the scale and identity of places, physical and psychological barriers, deconstructing not only the crossed urban fabrics and spaces, but also their community life too (in the *photograph*: the "Peripherique"—ring road of Paris); *photograph*: http://commons.wikimedia.org. **b** Consumption of urban public space, including in urban areas with high architectural and historic value, mostly central areas, by a tide of vehicles parked everywhere (in the *photograph*: Sf. Anton Place, Bucharest, 2008—rehabilitated in 2012). **c** Aggression of urban landscapes by conflicting relationships between cars and their other components: perception obstruction of the street's fronts bottom and especially those of valuable ones (in the *photograph*: Lipscani street before becoming pedestrian, Bucharest, 2008—rehabilitated in 2011). *Photographs* **b**, **c**: Mihaela Hermina Negulescu

ratio between satisfying mobility demand and respecting stringent need for protecting environmental quality of life.

The major objective in integrated mobility policies, land use and transport policies (LUT) is promoting the "gentle"*travels*[3] (non-motorized, bicycle, and pedestrian travels) *and clean public transport* (see Fig. 3), much less "invasive" and

[3] Environmentally friendly travels.

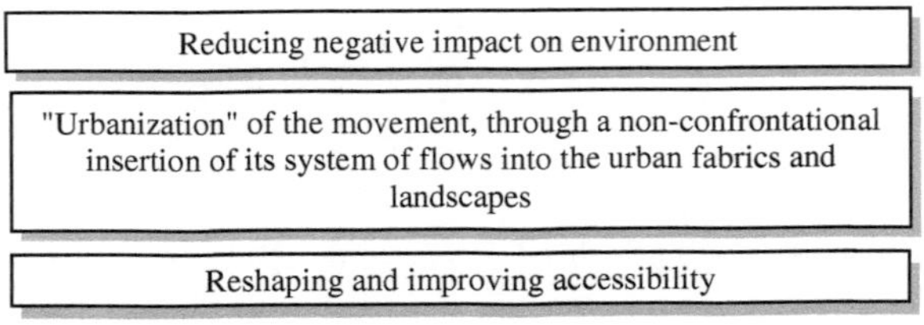

Fig. 2 General objectives of reshaping mobility (in light of sustainable development requirements) *source* Negulescu (2011)

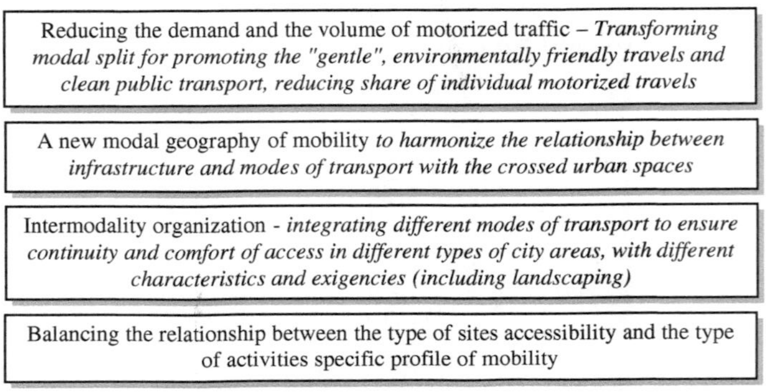

Fig. 3 The objectives of mobility policies; *source* Negulescu (2011a)

aggressive to space and cityscape; it is also a consequence of ecological and landscape protection requirements.

Bicycle and pedestrian policies take into account the fact that the simple implementation of specific infrastructure, although functional, is not a guarantee for success. It is shown that *changing mobility behavior, with a shift to non-motorized travels, depends heavily on the environmental and aesthetic quality of the route,* which significantly contributes to the improvement of travel activity comfort. The disadvantage of a longer travel time, at a lower speed, may be offset by the quality of "living" this time, in a beautiful movement environment, attractive and safe (see Fig. 4). That should turn travel as a "non-activity," a "dead-time" used to relate places, in a pleasant, relaxing, and leisure activity.

Landscape concerns related to mobility reshape are reflected in programmatic documents of spatial planning. Thus, the "New Charter of Athens"[4] provides that urban places and bicycle or pedestrian ways must be the concerns of composition and urban art," "Charter for the conservation of historic towns and urban areas"[5] states that "traffic inside a historic town or urban area must be controlled, and parking areas must be planned so that they do not damage the historic fabric or its environment," while "Guiding principles for Sustainable Spatial Development of

[4] European Council of Town Planners' Principles for Planning Cities, 1998.

[5] Adopted by INCOMES General Assembly in Washington D.C., October 1987.

Fig. 4 Adequate facilities for gentle travels (good, nice, sure, and correct organization for good functionality, for safety of movement, in different times of day and different weather conditions, in a pleasant ambiance and accessing attractive activities); *in the photograph* Aviatorilor Boulevard—a properly organized section of the infrastructure for cyclists, in Bucharest. *Photograph* Mihaela Negulescu (*source* Negulescu 2011c)

the European Continent"[6] recommends the control of the impact that Euro-corridors could have on the spatial development, environment, and landscape.

European Community documents on an emerging European mobility policy (Green Paper—Toward a new culture for urban mobility 2009) and E.C. Action Plan Urban on Mobility 2009–2012) emphasize the need to reshape mobility for protecting and improving the quality of living urban environment.

4 Landscape Rehabilitation in Urban Interventions Aiming at Mobility Reshaping

Most types of urban interventions undertaken in major European cities for reorganization of mobility aimed, by default, at rehabilitating the urban landscape, from "simple" esthetic upgrading, to complex recomposition, including by often

[6] CEMAT conference, 7–8 September 2000.

providing newly available public urban space, a valuable and limited resource which is, thus, object of a fierce competition whose winner was, for some decades, almost solely, the car.

4.1 The Transformation of Major Traffic Arteries in Boulevards

Changing the paradigm of human development and new mobility policies have caused a major shift in approaching and organization of the street, with the option to (re)harmonize the three of its hypostases: "traffic lane," "community area", and "urban landscape (Negulescu 2011b)." This implies that strict functional solution (considered the only one important, during the age of "car's sovereignty") has to be enriched with more serious concerns of composition and urban design. There is a return, a new emphasis on the concept of "boulevard," which reminds that street is (or should be) both a traffic infrastructure and a nice and comfortable public space for various activities, a place for walking and social interaction.

Transformation of major roads in boulevards is therefore a type of urban intervention, acknowledged, and promoted by all major cities, which is based on "democratization" of street principle and on the one of social and landscape street rehabilitation, considered as keystone of complex human settlements revival. Voluntary reduction in traffic lanes, increased allocation of roads space to gentle mobility infrastructure and vegetal elements, the rediscovery of urban art, the movement toward a new quality design for the urban furniture and for the exterior elements of utilities infrastructures, the conversion of some road intersections in places with central facilities such as squares, parks, transportation stations—all these are some recurrent measures in such urban operations.

An example is Barcelona which, in a complex "city project," has opted for the recomposition of some roads, especially in the outskirts, for the remodeling of the major streets as boulevards and for the rehabilitation of the urban places, gardens, and parks adjacent to them.

Paris also rehabilitated some of its boulevards, such as Boulevard Clichy which for many decades has been transformed into "traffic corridor" and recently has been reshaped through reducing the number of lanes for vehicles, increasing sidewalk surfaces, and creating a linear central public garden with bike lanes, rich vegetation and qualitative street furniture (see Fig. 5).

Bucharest presents also several important examples of boulevards whose landscape quality has risen in the last few years, by a correct arrangement and maintenance, through a complex management of the vegetation according to the season: Bvd Dorobanti, Bvd Kiseleff, Bvd Poligrafiei, Major Corvu St and lately Bvd Lascar Catargiu, which regained its central vegetal area, once gone (Fig. 6).

Fig. 5 Remodeling of Boulevard Clichy, in Paris. *Photograph* Mihaela Hermina Negulescu

Fig. 6 Boulevards in Bucharest—*left* Bvd. Dorobanţi; *right* Bvd. Poligrafiei; *photograph* Mihaela Hermina Negulescu

4.2 Development of the Clean Public Transport Networks with Dedicated Track as Structuring Axis of Complex Urban Interventions for Landscape Rehabilitation

In large cities and conurbations, the development of clean and structuring infrastructures for public transport networks with dedicated track—especially the tram—is often, beyond a way of improving the shape of mobility, the accessibility and urban cohesion, an opportunity and a central axis to extensive urban landscape rehabilitation. Thus, new implementations or improvements of urban rail infrastructure in European cities promote landscaping measures as: organizing grassy track, or with adjacent floral arrangements, or flanked by alignments of trees, the treatment of the transportation stations as beautiful architectural objects, even spectacular or iconic (sometimes with signatures of famous architects), special design of the technological objects for transportation (vehicles) and of their specific furniture (see Fig. 8). To those are often added landscape improvements in roadways and parking lots (see Fig. 7) in their area of influence.

Fig. 7 Landscaped parking (vs. common concreted parking lot) in Băneasa commercial cluster, Bucharest; *photograph* Mihaela Hermina Negulescu

An example considered a model of good practice is the major integrated urban intervention for remodeling mobility and rehabilitation of urban space, on a large city area of Strasbourg. As starting point of it is taken the traffic plan adopted in 1992, which proposed an ingenious travel system to meet both the need to reduce congestion, as well as that of improving quality of life and urban space, predominantly in the ancient city center, where a remarkable architectural heritage was degrading under the effect of increased pollution caused by motorized traffic. The new tramway was proposed as the center line of the whole urban development, being both: (1) structural element for urban renewal and cohesion, inter alia by rebuilding a proper relationship between the peripheral neighborhoods and the rest of the territory, (2) an effective tool to urban rehabilitation (see Fig. 8).

In the case of Paris, the objectives adopted with the Sustainable Development Plan (PADD)—to improve environmental quality and urban life—supported by the provisions of the PDP (Travel Plan of Paris) are materialized in one of the most important recent urban interventions: The reinsertion of the tram along the Marshals boulevards that constitute the inner ring of the outskirts of the city. This mode of transportation—ecological, clean, fast, regular, quiet, comfortable and safe—was chosen because it best serves the requirements related to improving the quality and convenience of transportation services, the accessibility, and because

Fig. 8 Landscaping potential of tramway infrastructure in cities. *Left* Strasbourg tramway—line A, Homme de Fer Station: infrastructure and technological objects with a refined design, well-integrated in the cityscape; http://commons.wikimedia.org. *Right* Tramway on the Boulevards des Marechaux, Paris—*photograph* Mihaela Hermina Negulescu

it meant, at the same time, an important opportunity for landscape rehabilitation of the involved boulevards. Thus, while operations of implanting the grassed rail track infrastructure were going on, a series of accompanying measures have been undertaken to improve the quality of public spaces along the route, according to the project of the landscape architect and town planner Antoine Grumbach. Vegetal elements were enriched by planting over 1,000 trees, the quality of pavement and urban furniture was improved, the number of motorized traffic lanes was reduced, and the amount of spaces for pedestrians and cyclists increased.

4.3 Rehabilitation of Urban Space and Landscape Around Transport Terminals and Intermodal Hubs with Attributes of Urban Centrality

In terms of urban planning, transport terminals and especially intermodal hubs—nodal traffic and transport infrastructure—are "major mobility equipments" for the cities, being at the same time cores of accessibility and centrality, impacting the proximate areas, and also affecting the attractiveness of the whole city. Therefore, organizing the intermodal transfer hubs is an important component of the (strategic) "projects for the city" and is often turntable of urban rehabilitation, renewal, and even restructure.

Urban developments for (re)organization of the intermodal hubs trigger in most cases the phenomena of positive dynamics to transformation and rehabilitation of their area of influence, grouping land valorization effects, proximity services amplification, and even processes of "invasion–succession" of functions with higher standard, all accompanied by the transformation of urban tissue and cityscape.

Fig. 9 The recently landscaped square in front of Strasbourg-Ville train station, in Strasbourg

Train stations, classic rail transport terminals, have become (or are becoming) *main intermodal hubs,* especially in big cities, with the particularity to constitute, together with their immediate neighborhoods, areas of contact, "gate areas" of cities, whose functionality, architectural and environmental qualities are representative for them and their community, contributing to their image and prestige. Therefore, metropolitan areas increased the concerns about esthetics and functional rehabilitation of stations' buildings, about their conversion into complex multimodal nuclei and, also, about the extensive *rehabilitation of their functional and landscape adjacent areas, of their influence area which is sometimes the whole neighborhood they are located in.* As opposed to the costly option to move the old stations outside cities, most of the time the variant of keeping them in the old quasi-central sites has been adopted, both as elements of architectural and historic heritage as well as major hubs for connecting cities to regional and continental flows.

There is often chosen the solution of creating an urban slab (vertical segregated and partially buried built complex) with an underground passage of the motorized traffic, on sequences of routes, and with the ground level dedicated to gentle movement and landscape outstanding facilities, adequate to the status of "rail gate" of the city (see Fig. 9).

4.4 The Creation of Urban Slabs, Coating or Burying Major Road Infrastructures

After a period of half a century, major road infrastructure, on ground or suspended, has disjointed cities, has cut them down into urban tissue fragments, and

Fig. 10 Coverage of A1 highway and its transformation in Boulevard Wilson, with a large median green space; *photograph* Mihaela Negulescu

has de-structured urban landscapes, the last two decades, brought in actuality the urban slab solution, not perfect, but sometimes preferable for reducing the conflictual and aggressive insertion of wide road structures in the urban areas they are serving or just passing through. This kind of urban planning releases *important amounts of urban space resources and creates tremendous opportunities for new and complex landscape designing* that may even recompose the identity of some neighborhoods.

Peripheral boulevard ("Periferique") coverage operations in Paris, in four sections (Porte de Vanves, Porte des Ternes, Porte de Champerret, Porte des Lillas), emerged as urban option from the requirements of noise and pollution reduction, from those of reducing the splitting effect that the great road infrastructure caused in Paris urban area, and from those of improving the quality of living environment.

Also in Paris, on the old royal road axis connecting the palace of Louvre and Saint-Denis Church, was built in 60s the A1 highway. The effect of physical barrier, the discomfort caused by traffic noise and pollution have led to the idea of covering it with a general slab and creating a promenade whose landscape planning was signed by Michel Corajoud. The huge green park, at the top level, is today a significant community place of the crossed neighborhoods and an urban landmark of Paris (see Fig. 10).

In Boston, recently was rerouted the great suspended highway (John F. Fitzgerald Expressway), on six lanes, through one of the most recent megaproject and most expensive urban developments in U.S.—*Central Artery/Tunnel Project,* known as "Big Dig." This huge urban intervention mainly aimed, inter alia, at burying its route through the historic central city, suffocated by chronic traffic congestion previously. Turning its 5.5 km of road mega-infrastructure into a tunnel was a difficult and spectacular engineering work that gave back to city inhabitants and to mild, gentle traffic a large urban land (105 ha), through an extensive functional and landscape recomposition of the public space, of which the most important are some plazas and urban parks (60 ha) and, mainly, the Rose Kennedy Greenway Boulevard, bordering a large linear park. The entire operation provided 3,300 newly planted trees.

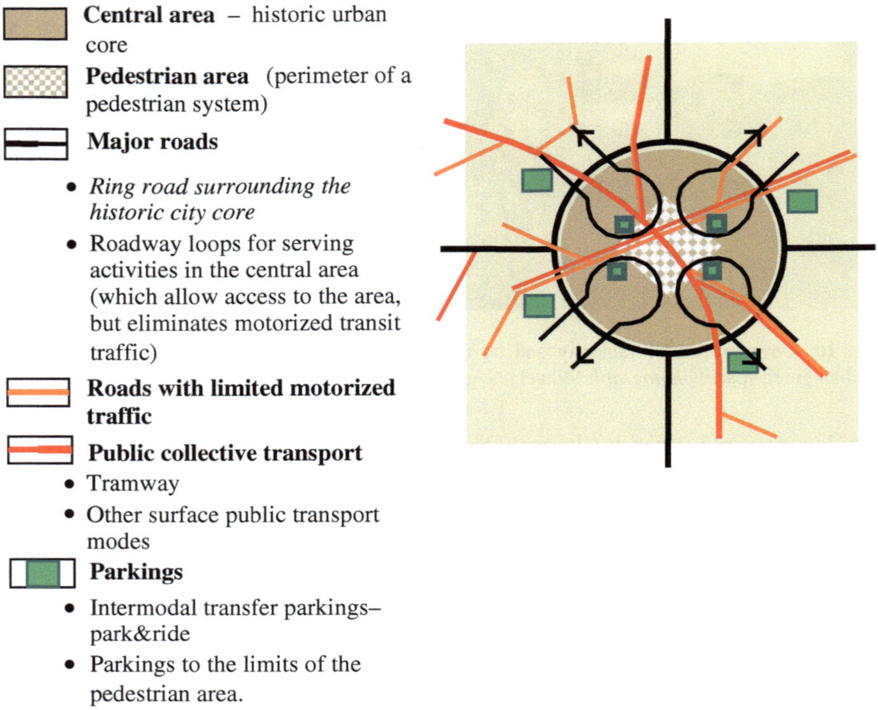

Fig. 11 Plan Strasse—mobility plan of reshaping mobility in the historic urban core of Strasbourg; author of the scheme: Mihaela Hermina Negulescu (*source* Negulescu 2011b)

4.5 Rehabilitation of Historic Central Areas Based on Reorganization of Mobility, Pedestrian Traffic, and Landscape Reshaping

A type of urban project quasi-generalized in cities with historic core is the rehabilitation of these central areas whose patrimonial, symbolic, representative, and (indirectly) economic value has to be recovered, protected, and improved. Degradation of buildings and ambiance is most often the consequence of previous policies, wrongfully oriented toward unlimited and unconditional satisfaction of individual motorized traffic demands, leading to pollution and to the "consumption" of large and valuable urban space. Therefore, such operations undertaken in most European countries—Germany, Belgium (i.e., historic center of Brussels), the Netherlands, Italy (i.e., renovation of historic centers in Bologna, Perugia), Spain (i.e., Ciudat Velle—historic center of Barcelona), Finland (i.e., downtown Kupio) etc.—imply a plan for mobility reorganization through reshaping of transport modal split, restructuring the traffic routes, creating infrastructures for pedestrians, cyclists and rollers, implementing automobility limitations, and providing good public transport services. Only under those circumstances, given the limited

Fig. 12 Place Kleber, Strasbourg; *left* in the period of "unconditional adapting city to automobility," *right* after rehabilitation urban interventions; *photograph* commons wikipedia.org

land resource, it is possible to commence the valuing of the built heritage through an improved and landscaped environment.

The large, integrated urban area developments, structured by the new tram line, in Strasbourg, among other objectives mentioned above, focused on rehabilitation of the historic city core (including Place Kleber—see Fig. 12), which, suffocated by traffic and pollution, was losing environmental quality and attractiveness to its tourists—the main advantages given by its exceptional urban heritage. There was an urge to adopt multiple, diverse, and interrelated measures that could be undertaken solely in the context of an intelligent modal split reshaping as that which was subject to the well-known Strasse Plan (see Fig. 11). Improving vegetal system, urban furniture, creation of a pedestrian streets system and a cyclists infrastructure, and revitalizing activities sensitive to the environment are the package of measures whose outcome is today valued and promoted as a model.

Similar operations, while on a smaller scale, were also undertaken in the Romanian cities with a valuable urban heritage: Sibiu, Braşov, Cluj, Baia Mare, etc. Rehabilitation of the historic center of Braşov, with measures for promoting pedestrianism in City Hall Market and on the Republic Street, was in the 1970s (with subsequent reviews) a much valued prime example of this kind (see Fig. 13).

The historic center of Sibiu, Cultural Capital of Europe in 2007, was also the subject of extensive rehabilitation interventions and valuation of the great historic built assets, which have been supported by an pedestrianization plan for the three old plazas: The Large Square, The Small Square, and Square Huet (see Fig. 14).

Fig. 13 City Hall Market, Braşov, Roumania; *left* in the 1970s, *right* in the present; *photograph* Mihaela Hermina Negulescu

Fig. 14 The pedestrianized Large Square, in Sibiu, Roumania; *photograph* Mihaela Negulescu

There are many other types of urban interventions, with a main objective of reshaping the urban mobility systems, which have an considerable impact on the landscape. Organizing the 30km/h areas, the car-free ones, or the shared-spaces streets, for instance, voluntarily aims to control or limit the motorized traffic, for the benefit of ambiance and quality of living.

5 Conclusions

It can be said, in conclusion, that *the vast majority of urban interventions for reshaping mobility comprises, among others, a subsidiary but explicit and sometimes spectacular component of landscape rehabilitation.* At the same time, the objective targeting the cityscape improvement, which often requires finding additional available land in the context of limited public resources, and as such, fiercely debated and usually under the aggressive monopoly of cars, cannot be achieved consistently other than by *designing and implementing integrated mobility policies* which, addressing accessibility requirements, have however to move the emphasis from quasi-sovereign automobile to the more environment friendly and less urban space consuming transport modes. *A better concern for the quality of transportation insertion in urban territories and spaces* and the reallocation of public space are essential premises for functional restructuring and for a better landscaping indispensable to improving the quality. of urban living.

Bibliography

Charter for the conservation of historic towns and urban areas, adopted by ICOMOS General Assembly in Washington D.C., Oct 1987

European Action Plan for Urban Mobility 2009–2012, Commission of the European Communities, Brussels, 30 Sept 2009

European Council of Town Planners' Principles for Planning Cities, 1998

European landscape convention, Council of Europe, Florence, 2000

Guiding principles for Sustainable Spatial Development of the European Continent, CEMAT conference, 7–8 Sep 2000

MODUR—Promoting sustainable mobility in Bucharest. Excellence research Program, Romanian Ministry of Education and Research, 2005–2008

Negulescu MH (2006) Ideological, theoretical and strategic evolutions in approaching urban mobility. In: proceedings of research in architecture and spatial planning national conference, U.A.U.I.M, Bucharest

Negulescu MH (2007) Innovation in mobility approaching and its impact on urban planning—recent issues in urban planning. In: Proceedings of the national conference in architecture and urbanism, UAUIM, Bucharest

Negulescu MH (2008a) Mobility and urban form—theoretical model of an urban planning with premises for sustainable mobility (re)shaping. Ph.D thesis, Bucharest

Negulescu MH (2008b) Theoretical benchmarks for spatial planning with premises for a sustainable mobility, in coordination with transports planning. In: Proceedings of transportation and land use interaction international conference, Polytechnica University, Bucharest

Negulescu MH (2011a) Mobility and urban form—theoretical aspects. Editura Universitară "Ion Mincu", Bucharest

Negulescu MH (2011b) Urban planning experiments of mobility sustainable reshaping, in practice. Editura Universitară "Ion Mincu", Bucharest

Negulescu MH (2011c) Transports in urban landscapes—problematic and best-practice. In: (ed) Landscape-architecture-technologies. Editor Universitară Ion Mincu, Bucharest

New Charter of Athens, European Council of Town Planners' Principles for Planning Cities, 1998

TERITRANS—Interdisciplinary proposals in transport system planning and in urban planning for sustainable development. "Excellence research" Program, Romanian Ministry of Education and Research, 2005–2007

Part VIII
Public Space: Exercise/Project/Intervention—Participation

Living Landscapes

Stephanie Brandt

Abstract An essay based on a series of panel discussions concerning the politics of public space, the potential of public art to engender creative participation in political and cultural life of cities and experimental or subversive urban practices, as organised and curated by Stephanie Brandt and Carol Mancke under the title of 'Living Landscapes' in autumn 2009—hosted at The Building Centre, Store Street, London and sponsored by the American Institute of Architecture (AIA). An attempt to reflect and document some of the most critical thoughts, future predictions and proposals regarding potential Living Landscapes as outlined within this context. Including contributions from Dr Malcolm Miles, Professor of Cultural Theory, the University of Plymouth, GB; Dr. Luis Arenas, Department of Philosophy, University of Zaragoza, Spain and Co-Founder of the research group (inter)section between Philosophy and Architecture; and Jean-François Prost, artist-architect, Montreal, Canada and initiator of the collaborative research platform *Adaptive Actions*. Living Landscapes investigates the notion of space as not simply being something defined by physical mass, but also as a repository of accumulated actions. Acknowledging the fact that contemporary cities are in constant flux and change, and communities no longer easily defined it addresses questions about how our landscapes are lived on, in and through. And how these landscapes and environments are revealed, imagined, animated and experienced by, in and through use and interaction? Following Malcolm Miles, public space has conventionally been regarded as the location of political life, or site of a public sphere. Today, when privatised spaces of consumption encroach on public space, it is necessary to defend public space as open to all elements of a free society. Sociologist Zygmunt

S. Brandt (✉)
Spacepilots, London, UK
e-mail: mail@stephaniebrandt.net
URL: http://www.spacepilots.net; http://www.stephaniebrandt.net

S. Brandt
Stephanie Brandt Studio, Sachsenring 28, 50677 Köln, Germany

C. Crăciun and M. Bostenaru Dan (eds.), *Planning and Designing Sustainable and Resilient Landscapes*, Springer Geography, DOI: 10.1007/978-94-017-8536-5_14,

Bauman goes even further, arguing that the task of critical theory now is to repopulate an effectively evacuated public realm.

Keywords Living landscapes • People and space • Space and politics • Spatial actions • Spatial experience

> *We will propose a possible conceptual topography of what is going to be called "fragile spaces". They are public spaces of disagreement and conflict, but also of mutual support. They are spaces, which do not represent but interpret: they interpret needs, deficiencies, shared hopes. ...*
>
> Luis Arenas

1 Introduction

The term 'affordance…suggests that the composition and layout of environment "affords" certain types of behaviour…Affordances are not inscribed in space but rather activated through the individual's sensory experiences, by moving through, touching, smelling, hearing or sight of objects and places…space is *performative* in that it affords certain practices and sensory experiences…the reconfiguring of public space involves a reconfiguration of affordances and resistances' (Degan 2003).

Living Landscapes has been initiated in 2009 by SPACEPILOTS in collaboration with architect/artist Carol Mancke. Living Landscapes strives to explore the idea of space as a repository of accumulated action. The interest is to unveil the 'affordances' of a place and to develop tools and methodologies that activate latent or luminal attributes of a place enhancing or enriching the way we perceive and use our surroundings—if only for a moment.

The project has three strands:

1. We started with an intervention in the public realm of an East London neighbourhood, which we realised in late 2009. http://spacepilots.net/projects/living-landscapes-01
2. A second strand of this project has been developed in form of a series of workshops with students in Japan, Bucharest and London.
3. As a third element of our collaboration, Stephanie Brandt and Carol Mancke have also set up a series of panel discussions hosted at The Building Centre, Store Street, London, exploring the matters and the alliances of: (1) *Public Space and Politics* and (2) *Art and Agencies*.

The following assembles some of the key thoughts, future predictions and proposals that came out of these debates. Aware of the limitations of this essay, yet aiming to build a clear case, the focus here is set on the talks and the following dialogue between Malcolm Miles and Luis Arenas during the first panel discussion at The Building Centre, Store Street.

2 Public Space and Politics

MM: Public space has conventionally been regarded as the location of political life, or site of a public sphere. Today, when privatised spaces of consumption encroach on public space, it is necessary to defend public space as open to all elements of a free society. Sociologist Zygmunt Bauman goes further, arguing that the task of critical theory now is to repopulate an effectively evacuated public realm. Yet public space does not equate with a public realm of political determination, nor a public sphere of political action. Today's consumer spectacles and spaces replace a previous generation of sites and statues by which the values by which a society's members should live were stated (…). Instead, for the most part, new and dissenting ideas take shape and circulate in transitional spaces, which are neither public nor domestic. This is not to refute the values articulated in differing ways by Bauman, but to argue for a more careful analysis of location, and for tactics outside the realm of design.

LA: (nodding)—Our lives are interwoven by such fragile threads. How can architects and city planners take account of them when they design public spaces? How can we possibly translate into volumes, shapes, programmes, uses, functions and appropriations the consciousness of this delicate matter on which our own existence relies? In other words: how can architecture give an answer to a situation, which is permanently *on tenterhooks*?

But, on the other hands, we may wonder whether this defence of the fragility of our public spaces is compatible with the traditional values of architecture. I realise that this defence could be controversial. Since Vitruvius, the Latin word *firmitas* resumes one of the three key aspects of architecture. *Firmitas*—which can be translated into English as firmness, solidity, durability or strength—has been one of the values of any construction throughout the history of the discipline. And is not *firmitas* the antithesis of fragility? Does not it seem as if there some sort of contradiction were lurking here? But I do not think so. In fact, the firmness evoked by Vitruvius is exactly the mark of fragility that was always hidden behind the concept of solidity claimed by architecture. Let me put it this way: our constructions have to be firm because they have to shelter the frailest thing of all: life.

So, from now on, the question will be this one: how can we design public (and even private) spaces that preserve and take care of this fragility? How can we think about this fragile space? Of course, as a philosopher, I am not the right person for this job. The task of the philosopher is simply to ask the questions, or to point out the problems, to those who are really engaged in the design of public spaces. However, I will try to suggest a few reflections on this topic.

First, I think we should not be tempted to try to represent this fragile space in our constructions. Architects should not evoke the fragility of our space through images and forms, as if they were conveying some sort of metaphysical thesis about the meaning (or lack of meaning) of our world. Architecture is not ontology (…). The urgent task here is not *how the world is*. This question is, of course, quite important. However, in my opinion, the basic question in architecture should be *how we want to feel about ourselves in the world.* As the Spanish philosopher Ortega y Gasset used to say, we, the human race, are not simply interested in *being in the world*, but in *well being* in the world.

If technology and arts are interested in how the world is, it is only because they want to *transform* the way the world is, according to our needs and wishes. Architecture should be ultimately committed to this political principle: it is our well being in the world, and not just our being here, that should guide architects.

MM: But, just taking Barcelona as an example, while the transformation of the city since the 1970s has certainly improved the sites of dwelling and sociation for many residents, improvement also has a history as a means of social control. In the nineteenth century, the improvement of urban spaces, with better conditions for living and working, and access to green spaces, was seen as lessening the likelihood of insurrection.

Perhaps the point is that a division of urban space into zones of private/domestic space and public space has not been—then or now—a means towards democracy, but to social control in the manner of liberal reformism (…).

There is a particular difficulty, (…): the sites of informal mixing and transitional activities on which everyday, tacit social formation rests are vital to the articulation of values and the shaping of a society by its members. In other words, if there is a public sphere in which this occurs, it is probably not located in public spaces but in the less definable spaces which are moulded more by use than by design.

LA: Architecture should act rather as a *mediator* in such conflicts (…). In other words, going back to idea of the fragile space—it does not *represent*: it should interpret: it interprets needs, lacks, wishes, etc. (…). A fragile space prefers the potential mistake and the risk of a *maybe* more than the sterile certainty of all that is already known.

I would like to imagine fragile spaces as fluid ones, flexible and *kind* to its users.

The Indo-European origin of the word *fragile* (the root '*bhreg*') confirms the suspicion that fragility is always threatened. Fragile things are always under the risk of being smashed, of being broken (in fact *to break* in English or *brechen* in German comes from that root: '*bhreg* = *fragile*'). The etymology of *fragile* also leads us to the concept of *fragment*, the parts from which it is possible to rebuild a totality. Hence, fragile spaces will always encompass dissonances, understood as a call for further reconstruction rather than as imperfections encouraging the destruction of their meaning.

A fragile space is, of course, a post-colonial space that avoids controlling and dominating its contents. It accepts interaction and dialogue. Using the metaphor that Edmund Husserl applied to philosophy, the architect will be a *public servant of humankind*, and not only a private servant at the beck of powerful people. Architects should recognise that 'being is expressed in many ways', as Aristotle remarked in his *Metaphysics*. Architects will accept that every culture has to, and tries to, modulate, in a provisional way, some profound truth as regards human mystery.

MM: There are times when a public space becomes the node of revolution and statues are pulled down, but this is rare. Mainly, public space is populated by representations of those to whom the mass public should look up—suitably on plinths.

LA: For this reason, fragile spaces also play with reflections, because reflected images lead us to a final question. What is the reflection and what is the reflected? Fragile spaces will be some sort of *trompe l'oeil* that will mislead or deceive our senses. They will fake conviction, even if they are just humble assumptions that, for the time being, deserve our trust. That is why, fragile spaces will favour alterity, without imposing it; they will invite subjects to rethink the coordinates of each situation, taking for granted the existence of alternative orders, of possible dispositions not yet foreseen. Understanding this subtle interplay of reflections, perspectives and interrogations can be a way of inviting us to think about reality in a way that does not allow itself to be grasped just from one perspective and, at the same time, in a way that allows every glance at it to be partial: it is nothing but a *fragment of meaning*.

Finally, fragile spaces will accept as a moral rule that, in order for life to continue on our planet, architecture and urbanism must be more welcoming and less brutal than they were in the past.

A fragile space should bear in mind that, in this new world, sometimes the most important things are the simplest things, the kind of thing that has become lost in the mists of time: an encounter, a conversation, a caress. One of the tasks of architecture should be to be faithful to the endless nuances of reality.

MM: Or, to put it more starkly, national governments are now largely outsourced providers of governmental services for transnational capital. Art, too, is co-opted to the agenda as Tate Modern's summer show in 2008, *Street Art*, demonstrated, graffiti, once read as evidence of antisocial behaviour, is now collectible. In these conditions, design has a limited role. Designers and urban planners may be able to construct democratic and participatory means of shaping cities, for instance in urban design action teams, but it remains against the grain outside the field of development architecture (in the majority world). Art may similarly be able to lend visibility to groups who are marginalised, or, perhaps more interestingly, it can critique the contradictions of the present order. At present, the most radical art tends either to address a specialist public—as in a billboard work by Mel Jordan and Andy Hewitt, in Sheffield in 2003, stating, 'The function of public art is to increase the value of private property', or it takes place in independently curated galleries. Clearly, the gallery does not house the public sphere, but neither did the coffee house. Both, however, may be locations of thinking towards its imagination and potential future creation.

Acknowledgments I would like to thank Dr. Luis Arenas and Dr. Malcolm Miles for their support in the publication of this essay. Further, I would like to thank Carol Mancke, Camilla Jarvis and Hyoung Shin for their collaboration in the Living Landscapes project at its various stages.

Reference

Degan M (2003) Regenerating public life? A sensory analysis of regenerated public places in El Raval Barcelona. In: Rugg J, Hinchcliffe D (ed) Advances in art and urban futures, vol 2. Recoveries and Reclamations Bristol: Intellect books, pp 21–22

Recondition the Urban Texture Through Public Interventions

Cristina Constantinescu and Mihaela Hărmănescu

Abstract During the time, every human society puts its own footprint on the space they are using (living space whether we mean public or private spaces), so when we are talking about the city, the thoughts lead to the heritage space and how we use it, daily living experiences into the space and its consequences. As a living entity, the city constantly is evolving and redefines itself through completion with the interventions, its identity constantly being a subject of the conditions born from pressure of the continuity—competitiveness relation. The effect of this relation is the transformation of urban texture, because today there are real issues faced by a number of cities that are loosing their controlled development and become more and more unsafe places, without identity. If we think this report, in the global market, in order to compete, cities are in search of the best congruent vision to take place in this competition and the regeneration of urban projects are most affectively and widely applied tool to achieve this target. The interventions are based on different attitudes and interests, but at the same time give rise to changes in long-term effect on the specificity of the urban tissue. Identifying the needs of a place whether it is about functions, activities, or the need for a physical form that meets its own minimum conditions to adapt to changes induced by the process of living, use, consumption, and redefining concern of society to create their living environment, with optimal space and distinctive as the society itself. Capacity of urban areas to be flexible and to change according to its needs is reflected in the quality of concept and from relations established through the project planning but also through their local integration in the general urban as involvement of the relations connectivity part in

C. Constantinescu (✉) · M. Hărmănescu
Faculty of Urban Planning, Department of Urban and Landscape Design, "Ion Mincu" University of Architecture and Urbanism, Academiei 18-20, 010014 Bucharest, Romania
e-mail: constcristina@yahoo.com

M. Hărmănescu
e-mail: mihaela.harmanescu@gmail.com

C. Crăciun and M. Bostenaru Dan (eds.), *Planning and Designing Sustainable and Resilient Landscapes*, Springer Geography, DOI: 10.1007/978-94-017-8536-5_15,

existing territories. Urban tissue as a product of the accumulation of various interventions as temporality, technique, scale, and attitude is the reflection process of continuously changing conditional framework such as the number of users, types of relations, the composite of segregation, social, ethnic, religious, political, and economical conditions. In this case, we are dealing with different approaches on urban tissue texture, with the development of ambiguous identity composition but which are the response of certain urban policies. What kind of mutation suffers the urban texture, how their consequences are affecting the urban landscape local identity and specificity, what is the relationship between space and all these characteristics? These are the questions that seek their answers in this paper. Listing some cases of existing projects, we want to reflect on the coherence of the approach and the way conception intervenes in the urban landscape. Hence, the paper deals with major interests of space syntax (cultural landscape, urbanism, public space) based on provocative analysis of the spatial relations between physical and abstract realities into the contemporary city and reveals the effects of the external pressure on the urban textures. When we are in a position to change a place, we should be aware that the final product will be more than its mere existence; it becomes part of local culture and accurate representation of the user.

Keywords Urban texture • Recondition • Culture • Identity • Landscape

1 Introduction

This argumentation is about the awareness of the role of continuous interventions on urban spaces, aiming permanent adaptation for the city's physical framework to current requirements. Article methodology is based on six major themes: theory of networks and defining the city as a flexible network of spaces (mosaic network) and relations; city as the result of the activities and actions that it hosts; analyzing the different typologies of urban interventions reveals us the multiple possibilities; urban landscape as a cultural landscape show the reality that the city is the "mirror" of the community that shapes it; public space project tools for transforming urban texture show us that these spaces deserve permanent attention of planners and always involve the cultural heritage of these places.

2 Theoretical Framework

The text purpose is to emphasize the intervention's importance inside the limit of the existing framework, interventions that aimed to improve urban environmental quality, contrary to the renewing trend of the urban tissue by peripheral extension. The tendency to support good relationships from the city with outside attraction is one goal of most cities. To be well connected to the territory, it may determine unexpected effect on the urban texture. Managing all the causes involved in the process of transforming city images depends on the ability to predict the transformation on the larger place than on a public square. The transformation that determines the new order (in cases when this kind of intervention is the only chance of the place to be reinstatement into the functional urban tissue) or the transformation that completes the existing urban texture are both the principal category of the intervention in the urban tissue.

A permanent reconsideration of urban tissue contributes to the responsive future of the city. Also it must be a permanent interest about this because the parameters are permanently changing and repetitive interventions are justified to be done in order to respond with new solutions to an old problem or to the new one.

More efficiently, the interventions are as if they are done to prevent the possible appearance of the new contradiction at the urban texture level.

Activities involving movement and movement attract attention, and for that reason, the mobility become in our days a new indispensable quality of the urban places. How all the urban tissue textures are organized as being able to manage all the flows of human interaction reflects the coherence of the urban intervention decisions.

Article suggests a reconsideration of the existing urban framework managing its transformation in order to prevent it from becoming unesthetic, inadaptable, and non-functional one.

Rebuilding the urban texture requires a careful reconsideration form of inherited values, which although altered by time or improper use, still have qualities that can relate to contemporary trends of development by integrating elements of local identity.

Through the theoretical approach are inserted and reviewed some of the famous urban interventions of the Hong Kong, Barcelona, Paris, and Tokyo, trying to define the main principle of the public interventions and their effects on the urban tissue texture.

3 City and its Formal Features: Tissue, the Reflection of the Activities—In or Out Pressure

The role of cities is complex, being defined largely on local culture and their ability to provide services targeting economic gains achieved at zonal and urban level, incomes that guarantee to the city a hierarchical position to local relationships at the regional, national, and global level. Cities ideally provide a set of services and products necessary for the citizens, and their involvement is characteristic in subsequent developments.

The indicators set which measuring the city performance is deduced from the transmitted effect to outside or to the territorial level. City performance could be measured by how much and how well it provides, considering the disposal resources and its ability to generate and support new relationships and exchanges that provide the framework and improve existing activities. Estimation quantification is especially difficult when the complex texture is based on a local-scale function, punctual, and by determining the area of influence in the first neighborhood and implicitly in the territory.

How it succeeds to identify the resources and uses it in order to ensure quality is widely recognized a permanent concern for the city to survive through strengthening relationships that ensure integration and cooperation with other areas and cities. Wide range of health services, education, infrastructure, administrative services, recreation, and environmental quality determine the extent of development. By providing these services, the city tries to reach in time common goals to respond from individual to stakeholders and establishing community-level social and economic priorities.

The needs of a city are not limited only to the exact area occupied; influences from outside of it are often determining factors in the functioning of urban areas.

There is the view that *the city does not grow in a single amorphous and undifferentiated space; it inextricably associates internal dynamics of an external response and external dynamics of an internal response* (Beaujeu-Garnier and Chabot 1963).

Quality of urban tissue texture results from its capacity to ensure specific urban activities, and for that, it must have the proper spatiality capable to respond at the necessities requirement. Proper spatiality means that the places are configured tacking count of the future facilities developing in the area but also follow the needs of connection of the area or place well with other part of the urban tissue texture. It must guarantee the proper performance of the activities provided by realistic strategy and by further interventions, meaning constant interventions that purpose to refresh the living environment. Not least important is a good government administration, to maintain a consistent minimum standard of quality that defines urban space.

The urban activities are directly related to the geographical position of the city that could admit or not some services. Inside, the city location and networking activities are influenced by the following: accessibility, infrastructure, source resources as a raw material, and human or technology workforce.

The physical characteristics defining the image of the city are completed by dynamics of activities involved in the urban area. There are two category of components that define the urban tissue typologies: physical elements that can be touch, seen having it's own formal proprieties and nonphysical elements that can be: smell, colors, quietness, noisiness, crowdies, air currents, active presence of the people, all that define the atmosphere of the place, what French people call *habitat*.

All that elements are more visible in the urban public spaces determined for the urban tissue evolution in time. Urban tissue texture is a result of lifestyle and activities of the citizens, but also programmatic interventions having the clear goals to make only the justified modification to improve citizen's life can determine a specific future behavior.

Because this entire context provides the human life, economic goals are strait connected to the city life and image, supposing that the urban tissues reflect all that. For the developing urban context meaning his activities and its formal shape, public intervention priorities always had also economic goal.

Today, increasing competition between cities by developing projects aimed at obtaining an attractive urban search is common use to benefit from economic consumption produced by the visitors.

With rapid urbanization, many cities are unable to absorb a high influx of population. Lack of shelters, basic services like water supply and waste collection, and jobs for these new residents have a significant impact on the city as a whole. Migration, both inside and outside the cities, is one of the main factors influencing and modeling urban tissue texture.

Urban system (all elements connected that define a city) is an imperfect system, and its imperfections are visible in the urban form, which contains at different levels layers and subsystems, constantly networking and subject to the same general conditioning and local–global constraints.

4 Typology of Urban Interventions Through Urban Tissue

Urban interventions are the punctual ones determined from private initiative to organize the private or semiprivate plots, but even if they are small interventions reporting to the scale of the city, these are visible from public urban space. These are interventions that if are dense in the city give an image of the people concerns.

Public interventions mean the interventions made for a mass of people not only for a punctual case of one private plot. Public interventions regard mostly the public areas and at urban scale can be interventions of reconstruction, to rehabilitate a quarter, a zone for the community, interventions to offer a new image of the place, and interventions to relocate activities or complete the existing activities with new ones or dissipate or complete clean an inappropriate form of the urban tissue. All those are changing the texture and the living shape of urban tissue. Public interventions in the urban tissue follow the different needs regarding the good and adaptive functionality but also provide the esthetic part of the image of the city (virtual image formed through the senses that perceive urban space; total information that is possible to receive through human senses such as sight, hearing, smell, and touch).

Cities are performing themselves operations of stratification, de-stratification, and re-stratification on the flows contained. Henri Lefebvre describes this organic esthetic of the flows of life in a city and its continuous restlessness and unpredictability as rhythms with as having a "maritime" quality: *There on the square, there is something maritime about the rhythms. Currents traverse the masses. Streams break off, which bring or take away new participants. Some of them go toward the jaws of the monster, which gobbles them down in order quite quickly to throw them back up. The tide invades the immense square, then withdraws: flux and reflux* (Lefebvre 2004).

Fig. 1 Plan and place atmosphere from the Fortified City of Kowloon, Fortified City Museum (Kowloon Walled City Park), Hong Kong (*source* author's photos Hărmănescu)

A radical example that can illustrate the reflux idea is Kowloon Walled City Park, Hong Kong (Fig. 1). The fortified city of Kowloon was once an enclave of crowded apartment buildings, linked by passages and dark alleys, a true heaven for the triads.

It is a specific case of economic and social dysfunctions that determines an urban intervention regarding a new type of organizing the area and also giving a new look. It was destroyed in the 1980s, and instead of it, there has been arranged a park with the same name and a museum as a memory of the place. Such radical surgery dissipation of a community came from the need to clear the area, from the inability of government control registration statistics for the people who use this place by excluding this enclave with high crime rate which relate to town and did not obey to fiscal duties. Has changed radically all textures that define the quarter, that in time it has become harmful for the city, being avoided by the tourists.

If the local organization of the users' activities define a lifestyle which for the city become an uncontrolled area that risk to enlarge and affect well function of the district could be a testament of local administration weakness.

Remaining as a testimony of a distinct type of living and showing the lived risk as an historical time interval, event has become in nowadays a testimony which reminds a lived stage of the city. Today, being accessible as an open public space become touristy reference. It is an example of how a rejected urban area becomes an incorporated part of the contemporaneous urban organization being a useful public urban space.

What was the strategy principal direction to renew the urban area texture that was an inaccessible part of the town was a social major problem of criminality that mutilates the urban texture unable to relate with other part of the city.

What was the solution followed to cancel all the inappropriate urban texture and dissipate the social mass resolved by reconfiguring a new form of area with new function useful for the city.

The complexity of tissue derives from the many forms of relations and from the diversity of contained points of interests and activities.

There are different locations; public interventions in urban texture could be in the following:

Central area interventions are sensitive at the historical heritage that defines this kind of places the form city central area.

In that kind of areas, the public interventions in the urban tissue try to reinforce the historical particularity of the places. The interventions are often at the small scale, regarding to regenerating a lost historical atmosphere, trying to use the existent old elements in this tissue, like the pedestrian and crossable gauge, the aspect of architectural elements, and the typical historical relation and their urban functionality and function.

An exemplary case is Paris Rive Gauche (Fig. 2), where we find a combination of contemporary and heritage reference consumption coexists in a harmonious and practical formula. It is about building heritage Grand Moulin of Paris that today became the university library, re-incorporating and accompanying across the new building of faculty of art and architecture. Assembly recognized under the generic name university develops around a central public space used by students for leisure hours. It is like a university courtyard, extroverted, directly related to the city (Constantinescu 2010).

In an urban public space project, first impact is the image and esthetics of proposed spatiality and its definition using an architectural framework.

One example, at territorial level, is Awaji-Yumebutai, Tadao Ando (Fig. 3) (Mostaedi 2002). The construction site at the start of the project was the scarred remains of a quarry, which had been gouged out and used in the construction of a huge landfill site for the Osaka Bay area. The project begun with the idea of taking land that had been destroyed through human intervention. Even before any architectural construction, the creation of small woodlands was the first step in

Fig. 2 Rive Gauche daylight use of urban spaces between new and old architectural heritage of the Paris (*source* author's photos Constantinescu)

an attempt toward the ideal. Development would not be premised on the destruction of nature, but rather the opposite: The project would restore nature to an area scared by past development and would recreate a new place in the area for people to gather and interact (Mostaedi 2002).

The attractiveness of a proposed space depends on the scale and volumetric new spatial form, but the most important part, even paradoxically is often visually indistinct, is technical drainage flow resolves, with appropriate accessibility and location in a coherent system functions included in the proposal.

Peripheral urban texture is permanently supposed to a pressure of residential or industrial areas development.

More common are cases when the desires of development produce new types of textures often exploits undeveloped areas on the periphery that generates a diffusion area with atypical structures that come to contrast with adjacent areas. Usually, this kind of extension does not define development; there are cases in which the new texture mimic an urban tissue. Making similar tissue copies form urban typologies, but in other conditions of other places, might produce an unexpected result often destroying the sense of the place. There are cases of developing residential quarters in Romania, applying the characteristic of American's residential development as area named Corbeanca residential, near by the capital (Fig. 4). That reveals that not only economic goals must be the only one of the developing mode of urban typology tissue.

What is dissipating the organization force of a city is the dilution and quantitative expansion without content that moves attention outside the urban tissue

Fig. 3 Awaji-Yumebutai, Tadao Ando (*source* author's photos Constantinescu)

Fig. 4 Overview on north part of Bucharest residential development (*source* author's photos Constantinescu)

and expands management areas of the city itself. The abandon of the urban tissue from urban central areas or major urban backbones contributes nearby of the urban uncontrolled extensions at the periphery, to a decreased environment quality of life and, therefore, to a heaviness ways of urban development.

Central core cohesion is determinate to sustain normal growth, radiating and attracting the development which can be achieved by isolated development spaces that break down the central cohesive force. But the coexistence between a representative central texture of a city and dissipated development areas into its urban landscape ensures the transfer and distribution of hierarchical interest into the amorphous mass of the city.

Although the initial term *urban reinforcement* was used often by geographers, *George Chabot used it in 1970 to Oprhrys, Strasbourg, defining as a "set of cities that control the life of a region," representing a "network of cities." Although the term "network of cities" was contested and proposed to use "urban reinforcement" term, both expressions are used today, they appear mostly to network technical references (within the meaning of municipal sanitary and infrastructure networks) that are reference to the habitat of the city (*Pumain et al. 2006).

From urban planner point of view, *urban reinforcement* seems correctly to be defined as a system of spaces that collaborate and support an evolution trend for the city. Urban tissue texture changes its consistency, and from the center area to the periphery, there are different types of spaces working together to define the specificities of the city.

Permanent link of the city with the neighboring territories is achieved by gradual transition from one typical texture of the city to feature texture organizations in the territory. Communication between the city and adjoining territories is materialized through the network of spaces designed to draw interest to visit and use it. Facilities and spaces from the public spaces of a city or from the territories administered by a community are substantial reserves that can sustain development priority, scheduled through projects which are focused to solve urgent needs.

5 Urban Landscape as Cultural Landscape

Urban landscape is a part of cultural landscape being involved with qualities and characteristics that define specificity resulted from the particularity of the local community life activities organization.

Almost all European cities are trying to define their own and unmistakable appearance, extracting spatial particularity resulted from resources and location.

There are landscaping designs of public spaces that invite us to use them daily. Barcelona is maybe the city that after Olympiads offers many spaces for the citizen's daily activities and encourages the usage of both categories of users: local people and foreign visitors. For that reason, the general atmosphere of the city is familiar for everyone (Fig. 5).

Urban tissue as part of the urban landscape is constantly a subject of dynamic change of uninterrupted life development footprint of its users. From the possible changes at urban level, those that are inevitably visible and have a major impact on awareness are those changes implemented by urban public space projects (market square profile, street, park, promenade area urban etc.).

Fig. 5 Barcelona daily experience—the human participation in some urban practice in many different ways, Miralles and Taglibue's Diagonal del Mar Park (*source* author's photos Hărmănescu)

Fig. 6 Barcelona daily experience—the human participation in some urban practice in many different ways define the cultural landscape—the sea front promenade in Barcelona (*source* author's photos Hărmănescu)

Integrated vision of urban space on the entire project is done by solving certain urban problems, improving the economic, physical, social, and environmental context—through the daily use of these spaces and events. Urban texture as a product of the accumulation of various interventions (temporality, technique, scale, and attitude) is the reflection process of continuously changing a framework of conditioning variables such as number of users, types of relationships, segregation, social mixed, ethnic, religious, political, and economic. The urban texture suffers progressive interventions, and how they affect local identity and specificity of the urban landscape, and the ratio of build space as negative and volumetric construction of architecture, is a question which seeks answers. The major interests of space syntax (cultural landscape, urbanism, public space) are based on provocative analysis: the spatial relationship between physical and abstract realities in and of the contemporary city. The urban experience has often been narrated as the reflection of an urban process between two plots: one as the materialization of planning codes and building ordinances geared toward maximum efficiency and realized through technical means, and the other as urban representation through the human eyes (Fig. 6).

We no longer see landscape *as separate from our everyday life, and, in fact, we believe that now as being part of a landscape, to develop our identity is a fundamental condition for our being in the world, in the most solemn sense of the word [...] Not only on its aspect or the conformity with a particular aesthetic ideal* Jackson (2003); we have to imagine the landscapes, but also by the way they meet our certain "existential" needs of the human being (existential needs which are, especially, emotional and social needs). *The contemporary urban environment is composed and recomposed by each individual every day around literal and virtual itineraries and not in relation to a fixed arrangement of places* (Pope 1996). Pope (1996). The life that humans generate marks the image of the city, and the quality of public spaces is a unit of measure for the quality of life in a city.

John Brunhes wrote *human activity Is reflected in the works* visible and tangible, *by roads and canals, among houses and cities, by clearing of the lands and through culture. [...]. There is still a human trace on the soil* (Brunhes 1912). The objective of the one, who studies the urban space, will be, in this respect, before any analysis and deciphering of human landscape work. The aspect of the landscape reflects this variable cultural attitude of the humanity toward the natural environment.

6 Public Space Projects as a Tool of Urban Texture Recovery

Reconditioning the urban texture starts from history, memory of the place, from how particular patterns and public spaces of a city were developed, from why they have been transformed, and how new forms were brought into place. To intervene, meaning to repair (recondition) the urban texture, means to win the necessary qualities of a competitive city.

Contemporary city requires maximum efficiency, so through changes, there are born new textures that are patterns of urban areas. Precisely, the dynamic transformations can help revive town identity by morphological restructuring, typology diversification, and through innovations of public spaces.

To understand the role of public spaces in urban regeneration through project, it is necessary to formulate the framework, which examines the interpretations implied by the concept of public space, but also the role of public space in the urban context. The urban public space implies a variety of spatial layouts, shapes, functions, sizes, and structures.

The public space projects determine the changing visible and consistent at the urban tissue because often involves a large or medium areas but every time there are visible because represents the support for the public active life of the city. Sometimes, the structural change into the urban texture involves the private propriety as well as public areas.

Reconsideration of urban public spaces as a priority in the process of renewing and improving the urban environment through interventions is due to the fact that public space in the city represents the legible part of the city.

People using spaces modify them according to their own reasons, but after a time or with generation's changing, same places become inappropriate for present daily life, and for that reason, the concept of renewing a place showing the fact that permanent changing of human structure enforce the places timely reconfiguration reusing goods or defining elements contained.

To improve public spaces, the council of United Nations Human Settlement Programme has adopted the first ever public space resolution (2011) which urges the development of *a policy approach on the role that place-making can play in meeting the challenges of our rapidly urbanizing world, to disseminate that policy and its results widely and to develop a plan for ensuring its application internationally*, where *the place-making is both an overarching idea and a hands-on tool for improving a neighborhood, city or region. It has the potential to be one of the most transformative ideas of this century...* HSP/GC/23/CRP.4/Rev.1. Because the importance of public spaces as living spaces, as instrument in urban recondition in line with the policy of spatial cohesion, as integrative force, as location factor in the competition between cities or as economic stimulus is gradually recognized and the revival of public spaces as a tool in urban development is gaining ground. The urban public spaces are crucial issue in social, cultural, political, environmental, and economic development. Public space is being used in multiple ways, often unconsciously for movement and traffic, for commerce, for communication, and for recreation. Urban recondition matrix made by the coexistence between two typologies—types of public spaces (street, place, park, etc.) and types of urban zones (center, first ring, second ring, periphery, exterior, etc.)—reveals one of the main characteristics of a public space—multifunctionality. Leaving from this feature and focusing on environmental evolution, which produces increasingly complex life forms over time, the public space, as a system, is a field of permanent genesis, and the constant flux of its subsystems is the means by which its social structure evolves with ever greater complexity. In evaluating thousands of public spaces around the world, *Project for Public Space Organization* has found that successful ones have four key qualities: They are accessible; people are engaged in activities there; the space is comfortable and has a good image; and finally, it is a sociable place, one where people meet each other and take people when they come to visit http://www.pps.org/articles/grplacefeat/. The Place Diagram was developed by the organization as a tool to help people in judging any place, good or bad. The center circle on the diagram is a specific place. The evaluation of a place is made according to four criteria in the red ring. In the ring outside these main criteria are a number of intuitive or qualitative aspects by which to judge a place; the next outer ring shows the quantitative aspects that can be measured by statistics or research.

Changes produced by proposing an urban public space project cut out a visible area from the urban texture with a strategic control and order at functional and esthetic level of urban texture.

The initiative of a unitary project that regulates an urban texture area prints a homogeneous uniform rule to adapt the intervention-affected properties to a set of principles and current needs at the city or territory. Reporting permanently to the urgent solution of urban texture spatiality highlights the vulnerability risk of future development, and how the new space created is able to adapt at the dynamic process of transformation induced effects everyday life. Occurrence interventions dictated by immediate needs often reveal cases of their limited survival, in the sense operation to parameters expected by the project objectives. Often urgent solutions rendered concerning present existing problems being a priority of the changes proposed in the project can blur historical context who reveal (even with a risk of subjective interpretation) a chronology of transformations on urban textures or on a place.

7 Culture as a Part of the Public Spaces Projects

Cultural events are actions contained in the urban environment. Public spaces are physical framework of human activities. Often, the cities increasingly compete by offering a varied calendar of events that will address to the different categories of users. A city can support such a strategy to promote the urban areas when the system is able to take streams and masses of people for which it is able to provide a good indicator and perception of public spaces in its context (examples are tourist cities).

Sometimes, the mixture determines curiosity to experience anything but usual. Often, public urban space morphology is the result determined by the cultural activities scheduled in the city as a reason for renewing and motivated intervention. *The proliferation of the use of public space increasingly gives form to society…a search for a combination of form and operational device that together create architectural space and quality. This combination provides a link between architectural space and urban, social dynamics* http://samueldavidbrown.wordpress.com/2010/12/24/chora-public-spaces/.

Commonly, interventions occur as an embellishment of the image of the place so that a minimal color and material changes in urban space produce effects that draw attention to a place. It is just a signal that the place needs to be changed. In other cases, that representing only that urban public space is used as advertising support with this role for a limited time sequence.

Habits of a community regarding some activities (sports, hobbies, local, daily habits, traditions) shape and reshape urban public space framework becoming typical for the community located.

The spaces are no longer seen as flats and linear but as a flow of matter-energy animated from within by self-organization processes said De Landa (2000). Furthermore, it is necessary to introduce the notion of self-organization, which became a crucial tool of comprehension for today's public space occupancy. While we comfortably allow environmental models to influence our perception of urban structure, eventually, we must translate patterns of human behavior into urban system and spaces.

By quoting Certeau (1980), on proposing a concrete strategy for resistance through daily life practice, we stated that people in the city can *counter and transform their urban space through individuals' participation and practice, for example, through walking*. The urban space changes with the position of the viewer or even better, the "flaneur" person leisurely strolling through its streets and became an experience. City context represents a three-dimensional medium of our life, involving the physical element but also the immaterial once.

8 Conclusions

1. Recondition the urban tissue requires permanent adaptation to field conditions and requirements of site users.
2. Intervention by organizing projects and redevelopment of urban public spaces is an initiative defined as visible and related community interests in or outside urban tissue.
3. Need regular intervention on urban tissue addresses two categories of functional and esthetic problems.
4. Evolution of the image and functioning of urban tissue requires a constant concern to identify new conditions and elements that predict the need for measures to change existing situations in a way that enables the city to become more porous, more connective, more sustainable, and more responsive to the landscape.
5. Urban habitat represents a part of cultural landscape that defines the characteristics of local humans' life, and the urban tissue is in a major part of it a result of the resident style of living.
6. Cultural events or daily urban activities impose the conditions of configuring the public space projects, involving the local particular elements from the place.

Bibliography

Beaujeu-Garnier J, Chabot G (1963) Traite de Geographie Urbaine. Librairie Armand Colin, Paris
Brunhes J (1912) La geographie humaine. F. Alcan, Paris
Certeau M (1980) L'Invention du quotidien, 1. Arts de faire et 2. Habiter, cuisiner, éd. établie et présentée par Luce Giard, Gallimard, Paris
Constantinescu C (2010) Argument volume, article: collective identity—the role of heritage values in the reactivation of contemporary public space
Pumain D, Paquot T, Kleinschmager R (2006) Dictionaire La ville et l'urbain, p 20
De Landa M (2000) Thousand years of nonlinear history. Zone Books, London
HSP/GC/23/CRP.4/Rev.1 Draft resolution on sustainable urban development through access to quality urban public spaces
http://www.pps.org/articles/grplacefeat/
http://samueldavidbrown.wordpress.com/2010/12/24/chora-public-spaces/
Jackson JB (2003) A la découverte du paysage vernaculaire. Actes Sud/ENSP, Arles, p 262
Lefebvre H (2004) Rhythmanalysis: space, time and everyday life. Continuum, London
Mostaedi A (2002) Awaji-Yumebutai, Architectural Design, Public Spaces, Links, p 156
Pope A (1996) Ladders. Princeton Architectural Press, New York, p 32

Media Landscape

Cristina Enache and Oana Andreea Căplescu

Abstract The landscape contains messages—about the living environment, climate, culture, society, and history. Reading the landscape reveals different information transmitted directly to the observer. The landscape exists in the eye of observer, and it must be seen to be understood and represented. This chapter debates how the messages are inserted in the urban landscape and the relationship developed between society, landscape, and information. The tendencies from the last decade manifest toward inserting the media message in the architectural object, of harmonizing the information transmitted through the urban landscape with the environment. Using urban digital mediatization, its organic integration into the urban landscape can help create a consciousness for the spirit of the place by communicating "stories"—historical, cultural, urban, and local, by organizing spontaneous urban events or attractive activities, by appealing to the community's latent knowledge. It produces practically a media landscape—a space of transmission, in which the individual is not only receiver (as it was happening until recently) but also a transmitter of information; the user also generates content.

Keywords Media architecture • Urban landscape • New media • Hybrid space • Media facade

C. Enache (✉)
Department of Urban and Landscape Design, Ion Mincu University of Architecture and Urbanism, Str. Academiei nr. 18-20, Sector 1, 010014 Bucharest, Romania
e-mail: cristinaienache@yahoo.com

O. A. Căplescu
Department of Introduction to Architectural Design, Ion Mincu University of Architecture and Urbanism, Str. Academiei nr. 18-20, Sector 1, 010014 Bucharest, Romania
e-mail: caplescu_andreea@yahoo.com

C. Crăciun and M. Bostenaru Dan (eds.), *Planning and Designing Sustainable and Resilient Landscapes*, Springer Geography, DOI: 10.1007/978-94-017-8536-5_16,

1 Introduction

The landscape is a medium. The medium is the message (McLuhan 1994). The message is received by the people. And the people rebuild the landscape.

The society belongs to the landscape that it inhabits and produces landscapes in its existential environment. The urban landscape is the spatial representation of the social culture and contains that specific information which was at the foundation of its genesis.

Over time, the landscape science has developed as a way of analysis, of understanding, of interpreting the human being in its relation to nature. Currently, the notion of landscape is linked with the social and cultural dynamics of the daily environment. We are in a special historical moment, a complex mixture of political, cultural, and social relations, whose characteristics strongly influence its development: the demographic explosion, the borderless urbanization, the depletion of natural resources, and the emergence of a mass environmentalist conscience.

The problems of the urban landscape arise in the context in which the city increasingly ceases to communicate with the environment. This rupture is encouraged by the unprecedented technological development, which satisfies most of the current needs of individual and society. Under these circumstances, the urban landscape needs to be the element of continuity in the city life. The urban landscape should provide the historical, cultural, and natural continuity, ensuring urban aesthetic values.

The landscape plays an important role in the urban restructuring process; it should also be seen beyond its aesthetics or decorative valences. The landscape offers above all the complex context, with strong influences on the social, cultural, and historical framework. It may become the element to ensure the continuity of space, the dialogue between new and old—between tradition and modernity— and between the anthropogenic and the natural framework.

Social and technological changes have transformed the urban landscape, and with the rapid growth in telecommunication and mobility, the public scene starts borrowing characteristics from the virtual world. The shift toward social media and the Web 2.0 paradigm (where the content is user generated) can now be seen in some public spaces where interactive media facades engage in a playful dialogue with their public, animating the environment and facilitating communication.

The focus of this chapter is to analyze how messages are brought into public space and inserted in the urban landscape, thus underlining what kind of relationships are between societies, landscape, and information and how they have developed and changed in the recent years.

In order to map the latest changes in these relationships, we first need a brief walk-through of the connections between society and landscape and between information and landscape while also analyzing the social dimension of space. Since one of the most prolific forms of media in public space is represented by publicized information, we will address publicity and advertising extensively focusing on issues of visual pollution and data overload. Another important aspect is the technology employed in the "medialization" of urban landscape, how this technology is embedded in the built environment and its effects. Taking a keen interest in the latest developments in outdoor media and their effects in public space, this chapter's conclusion aims to provide a base for understanding the potential of media in the urban landscape along with its social attributes.

2 About Communication: Information, Message, Communication Medium, and Society

2.1 *Information, Communication, and Society*

Communication is defined as transmission or exchange of data—whether it is written, verbally produced, through signs or behavior. Information is the sum of knowledge and data about a particular subject (be it a person, an object, or an event), and it is communicable knowledge about something (http://en.wiktionary.org/wiki/information).

The information has different meanings, depending on the various contexts where the term can be used (philosophy, science, or technology).[1]

The information lies at the foundation of the necessary communication process in the human society. Its transmission is a follow-up of social, political, economic, aesthetical, and cultural demands—and causes the construction of a (physical or virtual) communication and mobility system—which can be local, global, or universal. The communication can be seen as the main aspect of human evolution, the separation of the human species from the animal kingdom, as a characteristic of society—a need of integration and survival (the individual needs to express his/her belonging and needs education and culture in order to be able to live in a community).

Regardless of the scale on which it is produced, now the communication system has become mostly a technical product, generally constituted as an engine of technological development, being its motivation, following the quality and the efficiency, supporting a sum of economic issues (Enache 2013). Contemporary communication systems can be regarded as a by-product of industrialization.

[1] The term knows a large series of definitions. Generally is defined as "facts provided or learned about something or someone; what is conveyed or represented by a particular arrangement or sequence of things; data as processed, stored, or transmitted by a computer; a mathematical quantity expressing the probability of occurrence of a particular sequence of symbols, impulses, etc. as against that of alternative sequences" (http://www.oxforddictionaries.com). The term and problematics of information and communication have been studied by one of the authors of the article in her doctoral thesis. The urban landscape in the Informational Technology context (Cristina Enache 2010).

Media communication history offers a wide perspective on the ways of transmitting information, the mankind evolution being marked by three major communication revolutions (DeFleur and Dennis 1991), or "mediamorphoses" (Fidler 2004): the emergence of spoken language, of written language, and of digital language.

Information and communication advances involve changes in society's lifestyle, introducing new forms of interaction that have implications and consequences on the economic, cultural, social, and existential background of society.

In urban areas, the information is the essential element of orientation, identification, and structuring of space. This is directly perceptible through the senses (visual information), or deducted indirectly through other factors (indicators, spatial structure preceding or announcing a specific type of space, specific functionalities, etc.). The information is present in any urban area and determines its character and its quality. The way information is transmitted will modify the space taking into account the media communication typologies the society needs.

2.2 Society and Landscape

The landscape is the background of society's existence, being a witness of its history and its evolution. Moreover, the landscape is the reflection of a society, generating (on the individual, on social groups) filters that vary depending on culture and at the same time with it; the landscape is a result of human–natural interaction.

The urban landscape refers to the structure of the geographical area specific to a city and includes its biological, physical, and social components. It can be considered as a continuity of the containing space, expressing the structural characteristics of the environment. The urban landscape is the cultural expression of the society who dwells within and is the expression of its relationship with the environment. Therefore, the city represents the way the community has translated the physical and non-physical information existing in the landscape.

The morphology and evolution of the landscape describe the relationship between society and place, the way the man dwells the space. The link between a community and landscape is expressed through individuals' activities and the territorial behavior. The agrarian and industrial societies gave rise to specific landscapes, seen as a synthesis not only of the social or economic nature of its inhabitants, but also of historical and cultural foundations.

The powerful urban development from the second half of the twentieth century and the current globalization characteristics raise important issues on today's society new landscapes. The urban sprawl in the territory, the mobility growth, the information, and its communication produce specific typologies of landscapes. As the present society's individual develops various dimensions of his life in different places simultaneously, the experience of landscape is no longer tied to the place where we live, but to a number of territories which coexist when setting the itineraries.

An important number of issues emerged from the landscape analysis regarded as society's existence space. The social education causes the consolidation of community's concepts, of its identities and local values, related to the landscape's character and to the way it is perceived. The landscape can be shaped through social ideals, beliefs, or local values. The focus on the issues and on the potential of community highlights the importance of human resources as well as the harness of the identity, of the local environment uniqueness during the landscape structuring process following the evolution of society.

2.3 *The Social Dimension of the Space*

The space is the tool that architecture uses to assign meanings to the human existence. These meanings come from the values that we attribute to the shapes of the physical environment. As Norberg-Schulz says "the space is a necessary part of our existence" (Norberg-Schulz 2000) and consists of a set of objects with individual meanings, belonging to a wider context of multiple spaces and relations between objects. People behave and meditate in a complex space arranged for them and by them. But their behavior and their way of thinking depend more of how they perceive and represent the space.

Lynch (1966) theorizes the space awareness in Image of the City; he talks about the city's need to be readable, "the ease to recognize the parts and organize these in a coherent structure." People need to structure and identify the space for appropriating it; the cities need a powerful image—the product of immediate sensations and of past experience memory. Lynch (1966) defines the structure of an environment image as its identity, its structure, and its meaning, pointing five elements of physical shape that contribute to the image and identity of cities: "path" (along which the observer moves while noticing the elements placed on the way), "limits" (the linear space that defines the elements limit), "districts" (sections of the city whose boundaries have a specific identity), "nodes" (focal points, converging roads, parks, and squares), and "landmarks" (reference points).

In the context of informational era, the place becomes more than a space containing specificity and identity; it is necessary to adapt to the new dual structure, where the virtual adds a new dimension to the physical space. In today's society, more than ever, people need an urban landscape that would facilitate the creation of image, places with particular character, leading directions and reference points—memorable and distinct places. We have to underline the fact that the image is based on simple topological relations that man needs to build, appropriate, and identify its existential space.

Until 200 years ago, most people lived their entire life in one place and were developing a deep communication with it. At the debut of the twenty-first century, the motorization, the air travel, and the Internet have caused an unprecedented mobility. The sense of the place has changed its significance. The sense of the place promoted by communication technologies involves knowledge of the geographical diversity, generating interest for education, knowledge, and the accumulation of data.

The relationship between individual, space, and landscape acquires new meanings, and the idea that the landscape is intended to take over the problems of the contemporary city—spaces without identity, identifiable boundaries, to which the informational society adds its paradoxes. The urban public space has to satisfy the need of constant information, communication, and accessibility (consequence of mobility) and to become a node in the urban landscape—a network of memorable, identity places.

The spirit of place is a property—both inherent and emerging—that links the major structure of landscape to the anthropic life framework. It was born in the landscape, providing the spatial continuity that makes the individual aware of the environment of the place he lives in. It seems to be related to the territorial shapes and built forms, increased through accumulation of physical changes and associations. Identified in a particular anthropic space, it is in strong relationship with the cultural context, and it is shared at the community level through local history and geography of the place.

2.4 Information and Landscape

The urban space has always been under the influence of existing information from the surrounding landscape and simultaneously had to bear the modeling and development tendencies imposed by the location in the territory—crossing directions, relationships with the territorial system and the development of urban organism. It is an area where two types of information overlap—natural and anthropic that determine the dynamics of the landscape transformation. The information is in a permanent dialogue, in a process of reciprocal influence and modeling. If initially the cosmological data led to the creation of settlements, the way that communities have developed, brought interventions in the natural landscape adjusting it and have introduced new parameters in the understanding process of the space.

The electronic media transmit information around the globe in seconds, collapsing time and space led to the creation of the global village, where everything, no matter how exotic, can exist and seems now somehow familiar. McLuhan's argument is that—as communications media made environment changes, so the perception and the way of thinking are changed through electronic technologies that makes possible the global village, based on rigorous logic, assuming feelings, and emotions. Paradoxically, we live in an era in which the feelings are in the foreground, and the rationality in the background, because the media messages are primarily addressed to the senses, before the logic.

The recent increase in electronic media use coincides with the recent increase in cultural exchange that affected arts, literature, philosophy, science, geography, architecture, and urban planning. This change manifests itself in the rejection of modernist strategies, like those of replacing everything that is old with something new. Amazingly, the worldwide movement acts as a process of protecting heritage, a movement that did not exist before 1960 (towns of Italy, France, and Spain—almost abandoned 50 years ago—are reoccupied and revitalized due to strong identity).

Architects and planners like Jan Gehl or Kevin Lynch have dedicated their studies to define the urban form properties and what makes some streets or public spaces to be attractive, while others are abandoned. Observation of space involves selective accumulation of information, essential for its replication. Reconstruction of a place means to identify those elements, which give its individuality. A space is an accumulation of information that characterizes it. The information is therefore a tool through which a place can become recognizable and may obtain significance; its mediatization constitutes a necessary stage in the process of finding local identity.

Those features of space that can be experienced in everyday life are potential focal points in the process of identification, they are reference points for advertising messages or mental images thus becoming part of how the space is "communicated" inside and outside of the city. By intervening in urban areas, these images can be deliberately created, manipulated, and strengthened in significance.

The resulting conclusions concern the intimate connection created between elements belonging to history and tradition and elements of a very innovative typology that belong to the present time and, most importantly, to the future. The fact that the urban landscape is (or should be) strongly influenced by the major structure of the natural and built landscape relates to the classical concept of the spirit of the place. The Information Age in full ascension adds other parameters to the discussion: parameters related to mobility, to communication, and to information transmission. This is because the identity and character become major issues in the discussions about landscape—natural or anthropogenic.

3 The Publicized Information: Publicity

Since ancient times, the urban landscape was constituted as a mark of society, culture, living conditions, and technological developments. Full of meanings, symbols, and messages, the public space has always been a place for expression, interaction, and promotion.

The urban landscape speculates that role of information transmitter introduces messages directly addressed to the individual, for cultural, educational, commercial, and social purposes. Space, architecture, art, signage, billboards, and advertising posters, interactive displays, and media facades follow different purposes, integrating in the urban landscape through specific languages messages addressed to the space consumers.

The street publicity is the oldest form of advertising (billboards). Architecture has always been a form of expression—of function, of power (human or divine), of technology, and art introduces the aesthetic and cultural dimension in the urban space.

Fig. 1 Different types of messages transmission in urban landscape—**a** Notre Dame Cathedral, Gothic architecture—messages about religion, power, and building structure (building technology). *Photograph* C. Enache, 2005, **b** New York—street publicity. *Photograph* G. Zamfir-Khosravi, 2011

Messages are transmitted both from the gothic cathedral and from the electronic displays—the mode, the type, and the time of transmission are different (Fig. 1).

The twenty-first century's urban landscape is more than a morphological space—streets, plazas, and built space. The public space is a place of social, cultural, and informational interaction. The urban landscape facilitates the transmission of messages—through its spatial structure (monumentality, privacy), architecture (function, ambiance), advertising, media panels, and interactive facades (publicity, socialization, education, and communication).

The urban landscape has always combined the static dimension (topography, built frame) with the dynamic dimension (the movement—people, cars, light, climatic, and atmospheric condition). The new technologies combine the two dimensions, by integrating into the structure of spatial elements (facades, built elements) dynamic elements—like information transmission—mediatization, social interaction, and real-time information. The result is a hybrid space, the overlap between material and immaterial—between the physical space of the city and the virtual space of the message.

We can say that the landscape is a transmitting medium for the messages (gives and receives information). Clearly, there are two major types of information in the landscape: the natural environment information (primary, latent) and human environment information (anthropogenic). The urban landscape concentrates both the information of the place where the town was established with the information of the society that dwells the space. The way that society interprets the information is materialized in the built environment.

The communication of messages in the landscape takes two shapes of transmission: direct and processed.

Architecture, art, and the form of space—through the cultural filter—send messages—historical, symbolical, political, economic (through the architectural style, dimensions, localization, and materials) and—generally—reflects the image of the society who produced them. It forms a temporal "intermediation" (between the producing society and the society that uses them), a conceptual one (between the authors and the users), a communicational one (at morphological level, through the information complex contained by the built frame—scientific, artistic, economic, and motivational data).

The direct transmission of information in landscape is performed due to economic reasons (commercial—publicity), political reasons, or pure informative reasons (be it educational, cultural, or civic reasons).

The role of media images or advertising messages is to intermediate the communication of the spatial information to the individual; on the other hand, they represent the starting points in the process of awareness of space. These include, on the one hand, representations of real space that can be perceived and directly experienced (reproduced and communicated images), and on the other hand, spatial qualities that are indirectly perceptible (relations of territorial and functional structure with the space history through maps, photographs, diagrams, and interactive views).

The cities always offered advertising spaces—panels, posters, and signs—located in public spaces—like public transport stations, train stations, and intense traffic (pedestrian or cars) areas. The form and the localization have been dramatically changed over the last two decades, having a significant impact on the urban landscape (not only on its image, but also on the social, economic, and cultural issues).

The street publicity's history can be traced to ancient times. Along with the development of the first cities, as a result of agricultural revolution (the first technological revolution that determined the first forms of commerce and the division in social classes), thus appeared the first forms of urban landscapes. Place of social interaction, of exchange of goods and information, the urban space becomes also a place of messages—in Ancient Egypt (Taylor and Chang 1992)[2] the laws were displayed in the public space, to be acknowledged by citizens. The next important step was in the Middle Ages, along with Guttenberg printing press. In the debut period of Industrial Revolution, the first illustrated posters appeared for public display. Originally posted on the buildings, the panels start getting their own structures, depending on the space they were located (in relation to the development of cities caused by increasing mobility). During this period, the outdoor advertising had local character; the posters were designed by merchants to promote certain products, or for various public events (shows).

---

[2] Taylor and Chang (1992) The history of outdoor advertising. Regulation in the United States, Villanova University, p. 284. *The first recorded use of outdoor advertising occurred in Egypt, were notices of reward for the captures of runaway slaves were printed on papyrus and posted. The Egyptians were also known to have inscribed hieroglyphics on obelisks to direct travelers and merchants were known to carve sales messages into stone tablets which they placed along public roads.*

(a) (b)

Fig. 2 **a**, **b**. Visual pollution caused by the agglomeration of advertising messages—outdoor advertising in New York City. *Photograph* G. Zamfir-Khosravi, 2011

At the end of the nineteenth century, the first associations of advertisements promoting products from major manufacturers are established, in order to coordinate and to control the transmitted messages, putting problems of ethics into question. Besides, the classic posters appear fixed and mobile billboards and transportation shelters offering permanent space for advertising. The digital technology of the early 1990s introduced a new typology of outdoor advertising image. The drawn posters are replaced with printed banners, the facilitation of representation and production having strong implications, resulting in an explosion of outdoor advertising that flooded the city space. When this activity is controlled, the urban landscape is dynamic and attractive. There are cases when urban issues were solved by urban advertising, making them elements of attractiveness. But in most cases, the results are often negative, when the publicity is a visual and informative aggression, ending in some cases with the interdiction of outdoor advertising. (The space becomes a cluster of advertisements and posters promoting products and public events—Fig. 2 a, b.)

The introduction of screens and media facades—which besides the advertising function—also insert an interactive and informational dimension (information about weather, traffic, and the possibility of transmitting messages) and seems to be a solution of solving the problem of visual pollution generated by outdoor publicity, through harmonious integration of advertising into the urban space and sometimes in the architectural object.

4 Media Technologies in Urban Landscape: Social Interaction

The development of new technologies involves a completely different typology of expression in the landscape. Because the necessary infrastructure has not the amplitude from the industrial period, the devices are generally much smaller and

Fig. 3 Interactive urban presence—through new technologies applied to architecture—Water Cube—Beijing Aquatic Center, Beijing, PTW Architects and OVE ARUP. *Photograph* G. Zamfir-Khosravi, 2010

can be framed in architecture, thus developing a relationship with the environment. Even the urban landscape no longer suffers such dramatic changes; these changes are manifested more as a result of transformation in the social structures caused by new information, communication, and mobility technologies. The perception of space is affected, because passing through space is different now, requiring new elements, powerful nodes, and local urban identities.

The integration of information and communication technologies in architecture and urbanism is developing under the influence of software tools, of new ways of thinking and designing. New media changed the way we design, the way we think, and interpret information also affecting our expectation from architecture and urban planning (Căplescu 2011). The difficulty that arises is manifested in the need to rethink the methods inherited from the pre-information period, along with the spatial integration of new technologies. Beyond the theoretical strategies that are planned for an organic and controllable development of the city, there are manifestations of new technologies in the urban form, results of the society's need for communication—best expressed at the level of public space. Integrated into architecture, the communication technologies applications become part of urban culture, giving identity, and particularity to the urban space (Fig. 3).

The urban object receives a great importance in the new landscape, whether it is a public space, an architectural object, or an element of urban art. In this context, arising elements or systems of functional elements required by the new demands resulted from technological developments that shape the contemporary landscape—urban, suburban, rural, or natural. The intermodal nodes, generally located outside the cities, coherently connected to territory, ensuring the transition between different modes of transportation and forming real centers of interest, by polarization of other functions—complementary or compatible, become functional and volumetric accents in the environment.

Media facades represent a phenomenon affecting the urban area and the individual—visitor. These transform static architectural environment in a dynamic communicative space (Fig. 4).

The buildings turn into a living organism and begin to interact through the informational program, thus becoming a permanent presence in the urban landscape, an

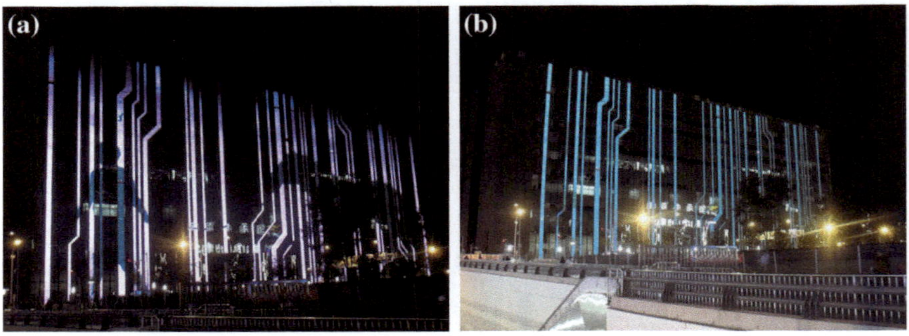

Fig. 4 **a, b** Digital Building, Olympic Park, Beijing, Studio Pei-Zhu—on the facades are transmitted images of various traditional legends. *Photograph* G. Zamfir-Khosravi, 2010

integrated part of the everyday life's space. Through this, a relationship is established between the individual and the urban environment, harmonizing the needs of the users in communication with the site and capitalizing the identity of place. It is in fact the birth of an urban environment adapted to the universe strongly influenced by Internet and other media. The rebirth of public culture is brought by street art, with media facades and with interactive urban landscape.

Digital media can be used for more than just advertising. It can communicate civic or educational messages and useful community information and could attract attention to public urban issues. It can bring attractiveness in a public space, may provide social interaction, and can help create a public conscience and raise awareness.

The communication in the physical space using the digital technologies gives rise to the media landscape—an accumulation of communicational infrastructures, of spatial perception elements, of individuals—receivers and space consumers, subliminal messages following the mediatization purpose. Secondary issues target social interaction, spatial appropriation, creating landmarks, and urban identity.

The communicational infrastructures in the landscape can be:

- Frameworks for outdoor advertising.
- Technical infrastructure for communication (antennas, cables).
- Visual (poster, commercial, and digital displays) or virtual messages (suggested or induced through the image or the landscape).

The mediatization in the urban landscape may have the following effects:

- Negative—excessive outdoor advertising (aggressive, assaulting pedestrians, and the architectural framework, leading to spatial alienation).
- Positive—by spatial integration of the message—by leveraging the physical urban heritage (architectural, artistic, and historical), increasing the attractiveness, introducing landmarks, developing a spatial appropriation feeling, socializing activities, and enhancing space consumption.

The unprecedented growth of mediatization in the last 20 years has implications for democratization and public discourse. By promoting pictures and information about the contemporary world, the media plays an important role in shaping social consciousness. The mediatization provides a link between people, different degrees of closeness and social solidarity.

The hybrid space resulted from the interaction between real and virtual is called by Scott McQuire "media city" (McQuire 2008).[3] He says that the migration of the electronic displays in the urban landscape has become one of the most visible tendencies of the contemporary urbanism. Along with the development of global networks and of mobile media, they have a deep impact on the urban space (McQuire 2006).[4] As Paul Virilio says, the architectural prototype of "media building" appears—one that hosts information rather than dwelling (Virilio 2001).

Digital media contributes to increase the feeling of appropriation (de Lange and de Waal 2011), meaning by appropriation the space assimilation by citizens, involving them in solving urban problems—either global or local. The direct and visual communications, the simplified messages, and the processed information—accessible to anyone—have a significant impact on the individual.[5]

It outlines an interaction environment, beyond the economic, political, and religious influences. The access in the public sphere is open to everyone. Promoting the information and images shape the social consciousness, the collective life, and the urban history; the individual absorbs information that becomes part of the collective culture.

5 Conclusions

The relation between form and context has changed. Because it involves adaptation, and the needs and the aspirations of the society are now different, the conception of shape has changed its parameters; therefore, its acceptance by the context determines other ways of interaction between the natural landscape and

---

[3] McQuire (2008) Media City, Sage Publications Ltd, 2008 ISBN 1412907934. *Rather than treating media as something separate from the city—the medium which 'represents' urban phenomena by turning it into an image—I argue that the spatial experience of modern social life emerges through a complex process of co-constitution between architectural structures and urban territories, social practices and media feedback.*

[4] McQuire (2006) The politics of urban space in the media city, available online http://firstmonday.org/htbin/cgiwrap/bin/ojs/index.php/fm/article/view/1544/1459. *Electronic screens do not form part of a building's memory in the way frescoes or stained glass windows could; rather their restless constantly changing imagery contribute to a dematerialization of architecture, a sense of ephemerality which is pervasive in twenty-first century urbanism.*

[5] de Lange and de Waal (2011) Ownership in the hybrid city, Virtueel Platform, available online http://virtueelplatform.nl/english/publications, p. 7—*information technologies play a major role in the urban experience, saturated as it is by mobile communication, Wi-Fi, GPS navigation RFID cards, camera surveillance, urban screens in the public space, and so on. For a long time the domain of digital media was viewed as virtual, as something separate from physical reality. But now these two worlds are tightly interwoven. The contemporary city is a hybrid city with physical and digital infrastructures, services and processes at all levels.*

the anthropic landscape. The connection between man and place has changed its nature—more mobile, the man no longer develops deep relations, as in the old society. The ITC implies territorial delocalization. The place becomes an accessory, a support for activities, and its main characteristic being connectivity (translated through the accessibility to network, to the needed infrastructure).

The current understanding of the city, as a response—on the one hand—to the need of technical systems articulation, and on the other hand—to satisfying the social needs, involves an analysis that includes developing networks whose territorial reference is the entire planet. The multimedia network ignores the physical notion of the city. The reorganization of territory and urban space is made through new infrastructures, satellites, and optical fibers, by modifying the landscapes of our cities. New forms of territorial delocalization are added to the traditional urban problems. We are not able to master the rupture between space and time. The concepts of multi-territoriality and de-territoriality lead to the loss of city's memory.

The urban memory must be strong enough to determine changes and impose the structuring of new urban entities. The existing urban frame—sometimes shapeless and incoherent—needs to refind its identity, by re-inventing itself. The role of image and architectural object's significance will fundamentally be to restore readability of urban space. The spatial coherence is an essential condition for building a local character. In the context of the information and media communication technology, the urban character is identifiable through the concept of "brand"—being a recognizable concept in the globalization era.

The city has developed increasingly close relationships with digital communication technology and electronic transmission of information. It produces an overlap between physical and virtual world, the landscape becoming the message transmission environment.

The urban landscape is the public scene where the urban culture and the need for social belonging are the main "characters" of the play, and where media technologies have an important implication through the information submitted, following the ownership of space, identifying the spirit of the place (by historical narrations, local stories, anecdotes). These determine new urban roles for involved citizens, in opposition to the tendency of growing individualization of the Information Society.

We have shown that information and its communication is one of the main players in the urban landscape, reflecting and shaping its society. Due to the technological changes, the digital revolution, and the expansion of telecommunication, a paradigm shift has occurred, bringing new meanings and relationships to urban public space, its identity and its inhabitants. The increased mobility along with the added layer of virtual spaces and localization technologies redefine notions of place and identity. The need of specificity and customization has become a focal point in the developing strategies for urban spaces and urban landscape design. In resonance with the changes in communication technologies and the emergence of social networks, a new layer of digital media infused the city. The media landscape has become an element of identity, able to link people, places, history, and cultures. While the risk of information overflow and oversaturation with stimuli that alienates its users is ever-so present, we also underlined that media landscapes can become catalysts for renewal, re-appropriation of urban space, cultural promoters, and enable social interaction, thus bringing a new dimension to the relationship between society and urban space in the twenty-first century.

The media landscape is both an answer and an effect to the changes brought by the digital revolution in both society and its built environment, and it is a new stage in the shaping of our cities.

References

Căplescu OA (2011) Implicaţiile noilor media în arhitectură. Doctoral thesis, Ion Mincu University of Architecture and Urbanism, Bucharest, Romania

DeFleur MF, Dennis EE (1991) Understanding mass communication. Houghton Mifflinm, Boston

Enache C (2010) Peisajul urban in contextul tehnolgiei informationale. Doctoral Thesis, Ion Mincu University of Architecture and Urbanism, Bucharest, Romania

Enache C (2013) Oraşul şi societatea urbană în era informaţiei. Editura Universitară Ion Mincu, Bucharest

Fidler R (2004) Mediamorphosis. Să înţelegem noile media. Idea Design & Print, Cluj

de Lange M, de Waal M (2011) Ownership in the hybrid city, Virtueel Platform. Available online http://virtueelplatform.nl/english/publications

Lynch K (1966) The image of the city. The MIT Press, Cambridge

McLuhan M (1994) Understanding media: the extension of man. MIT Press, Cambridge

McQuire S (2008) Media city. Sage Publications Ltd., Thousand Oaks

McQuire S (2006) The politics of urban space in the media city. Available online http://firstmonday.org/htbin/cgiwrap/bin/ojs/index.php/fm/article/view/1544/1459

Norberg-Schulz C (2000) Architecture: presence, language, place. Skira Editore, Milan

Taylor Ch, Chang W (1992) The history of outdoor advertising. Regulation in the United States. Villanova University, Villanova

Virilio P (2001) Spaţiul Critic. Idea Design & Print, Bucharest

Part IX
Conclusion: Territory/Detail/Resilience—Sustainability

Planning and Designing Sustainable and Resilient Landscapes: Conclusion and Recommendations for Further Work

Maria Bostenaru Dan

Abstract The book tackles a topic very much neglected in current research: it is not a book about visualisation of landscape which is more spread, but about design issues, at both scales of landscape architecture and landscape planning. This book deals with planning issues in landscape architecture, which start at the evaluation of the existing fabric, its history and memory, approached and conserved through photography, film and scenographic installations, a way in which the archetypes can be investigated, be it industrial derelict sites or already green spaces and cultural landscapes. It provides approaches to intervention, through rehabilitation and upgrade, eventually in participative manner. To such evaluation and promotion, a couple of disciplines can contribute, such as history of art, geography and communication science, and of course (landscape) architecture. The field of landscape architecture reunites points of view from such different disciplines with a view to an active approach, a contemporary intervention or conservation. The book presents case studies from several European countries (Romania, Austria, the United Kingdom, Germany), mostly for large landscape in the outskirts of the cities and in the parks.

Keywords Sustainable • Resilient • Planning • Design • Intervention • Conservation • Discipline • Garden • Photography • GIS • Map • Street • Agriculture • Public space • Cultural landscape • Memory • Interview • Digital • Media • Forest

1 Introduction

This volume is produced at the 10-year anniversary of landscape, in two curricula "Landscape design and planning" and "Landscape and territory", both led by the first editor, at the "Ion Mincu" University of Architecture and Urbanism.

M. Bostenaru Dan (✉)
Urban and Landscape Design Department and Centre for Architectural and Urban Studies, "Ion Mincu" University of Architecture and Urbanism, Academiei 18–20, 010014 Bucharest, Romania
e-mail: maria.bostenaru@iaim.ro

C. Crăciun and M. Bostenaru Dan (eds.), *Planning and Designing Sustainable and Resilient Landscapes*, Springer Geography, DOI: 10.1007/978-94-017-8536-5_17,

We aimed to make a book on landscape design seen from the point of view of more disciplines. Initially submitted to "Urban and landscape perspectives", this book, which became a volume in Springer Geography, builds thus on approaches already approached there on "Transdisciplinary knowledge production in architecture and urbanism" (Doucet and Jansens 2011). We look instead to multi-, inter- and transdisciplinary approaches in landscape design. Among those represented are architecture, urban planning, geography, history of art, communication science and ecology. We do not approach transdisciplinarity per se, but the points of view of more disciplines are applied to the same investigated subject. The aim of the book, conservation and intervention, points at the fact that designed landscapes are heritage, a heritage which changes over time. We approached how we see this at the beginning of the twenty-first century, looking back to the past and to memory, which is an important issue in the book. We will detail how the memory conserved, different from the reality which faces either conservation or intervention, is approached by different disciplines in these short conclusions of the book.

Conservation and intervention are both looked at with a view back to history or in contemporary development. We did not stop at theory, but provided case studies with the aim to highlight the contribution of different disciplines. As we will see, common ground can be seen either in the issue dealt with or, even more relevant, in the instruments (such as photography) which are employed for specific methods.

The book aims at a cooperative approach of specialists from more disciplines, as we try to draw the conclusions, since phases of intervention or conservation of landscape values are from the identification of heritage values till communication of these to the public, but not only through participative means. Another means of communication is the visual communication of memory through different kind of images and visualisation means. The landscape architect is understood as moderator between these professions in the process of decision making (s)he makes for design, a different view from considering the landscape architect a creator competing with other professions (Deming and Swaffield 2011). This is what we meant to reflect by depicting all views of these disciplines. The input from the different disciplines helps in the design process, by valuating the existing situation, the basis for all design and planning decisions for a future development, or by including, through participation, the user in design and planning, for which it helps how the landscape is perceived.

Although dealing with designed landscapes, the focus being on the design side of the profession, and not on the analysis or visualisation, the papers included are research based. Most of them investigate developments in macroscale landscape (agricultural landscape, rural landscape, the landscape of whole counties when studying forest development—which could be silvicultural landscape) and only a few researches by design at microscale. It will be the subject of a future book to deal with research by design at macroscale, that is, the development of strategies. It may be, however, mentioned that another book of us presents such a strategy: the development of Brăila County landscape, a project lead by Crăciun, the main editor of this book (Crăciun 2014).

We try to situate designed landscapes between landscape architecture and landscape urbanism/landscape planning through these differences in scale, since design (in German well expressed with the term "Entwurf" different from "Projekt") applies to all scales, design and planning alike.

Although wishing to provide a guide for landscape design, our book is by no means a manual, but rather a cross-vision of what is happening in the field, and we try to issue recommendations on issues to which researcher should be further dedicated here. Aimed at landscape design, it also not a reader stopping at theory.

The next volume, which might be in the "Urban and landscape perspectives" series, will deal with designed landscapes as well, but from the point of view of research by design: development of strategies. It will be research by design at the scale of the projects investigated here, with methods shown in some papers at a smaller scale. We will include also an aspect which was only marginally included in this volume, that of visualisation, which we detail in these conclusions.

2 Approach of the Contributors to the Relationship Between the Conservation and Intervention in Landscape Design

The case studies presented in the book exhibit a wide variety through the geographical coverage, theoretical and methodological approaches, standpoint influencing the focus of analyses, and scale. Nevertheless, several common elements can easily be detected in more, if not all the paper presented. Some relate to the research approach, some to the subject of the paper and some to the relationship with the subject of the book.

The first element emphasised by several authors is the importance of multi- and transdisciplinary approaches to landscape design. Petrişor defines the term landscape from this multidiscipline perspective, as an introduction to his approach. Also, Popa introduces definitions of landscape. The paper of Stan on urban culture is an extended definition of the cultural landscape. In the other paper, on peripheries, Stan introduces a new definition of the urban landscape planner, connected to the new term of "landscape urbanism", which is not mentioned, however. With this, we situate us in the book related to planning rather than architecture, to landscape urbanism, rather than landscape architecture, which is, however, as we will see, not followed in scale by all papers.

Some papers deal with the description of approaches, such as Popa or Stan, and some others approach case studies.

The case studies differ in scale. As such, the paper by Petrişor focuses on the planning scale, that of a county. Hărmănescu also works at a similar scale, but uses instead the historical delimitations, of the so-called 'country', a region. Other approaches are rather dedicated to landscape architecture, at the scale of gardens. The scale in between is the scale of the perceivable large landscape artefact, such as the agricultural field (Popa) or the periphery (Stan). To the scale of the gardens, we should subscribe the public space and the approaches to participation highlighted by some papers (Stan, Brandt, Hărmănescu, Constantinescu and Hărmănescu). The garden is seen in relationship to architecture (Meder and Krippner, Bostenaru). Also, some papers deal with a single case study (Covasna County, Ţara Haţegului, Scotland as territories, or Vienna as city), while other papers exemplify their theories with multiple case studies creating a regional spread of their concepts (the green wall, the pedestrian transformation, the media facade).

The relationship between planning and participation has also been explored by several authors (Brandt, Hărmănescu, Stan). Also, the paper by Negulescu is in the same line of the critique of functionalism which participation expresses, this time of the zoning regarding circulation.

The papers in this book can be divided into two broad categories: those dealing with technical issues and those dealing with creative spatial design. The paper of Petrişor looks at the quantitatively explorable analysis of landscape, as support for future strategies. Such explorative analysis could be also carried out by remote sensing, not only by GIS. And this is how an important part of the work is introduced: the photographic analysis. However, the papers in the volume deal with artistic, rather than documentary photography, as element for representing and interpreting the landscape. A first paper is that by Popa. It focuses on contemporary photography or a series of historic photographs until the contemporary state as exploration tool in landscape analysis, with a clear design outcome envisaged, namely how to revive historical landscapes. Popa introduces the link between the exploration by photography and the innovative way of analysing a site promoted by Kevin Lynch (1960) in "The image of the city". Unlike traditional planning that can be supported by GIS, this kind of research proposes constitutive parts of the landscape as it is perceived and can be represented in plan (zone, trace, boundary, landmark, node) which makes the transition to an innovative way of representing in new media as proposed in the paper on urban routes in an urban landscape by Bostenaru and Dill. Photography can be used in the later in order to identify the landmarks which are the nodes of proposed traces by different means. It is the same approach to the language of a landscape, seen as either "semiotics" (Popa) or "space syntax" (Bostenaru and Dill). The other article by Popa deepens the analysis of what is represented in photography, by establishing a code to interpret agricultural landscape, which she deepened in her photography paper. Herewith, we have a further term to "semiotics" and "space syntax", that of "code". For this, she distinguishes between what we would call in urban analysis container and contents. Popa differentiates between "surface" and "programme" which correspond to, respectively, surface and manifestation, or if we were to return to the container and contents definition of the urban organism to space and activity if we see it as a system (cybernetic approach). The urban organism as a complex system has two morphological elements which correspond to two interconditioned layers (Sandu 1988):

- At the phenomenological level, we distinguish the content of the system, called "urban life". The "urban life" consists of all urban activities together with a precise location. These localised urban activities form the urban functional structure. Going to the essence of these urban activities as part of a system, the urban functional structure becomes "the way of life". This way the spatial-functional layer is formed;
- At the physical level, we distinguish the container of the system, called "urban frame". The "urban frame" consists of the totality of the spaces corresponding to the site where they take place. All spaces together, corresponding to the location where urban activities take place, form in functional spatial co-operation the urban spatial structure. Going to essence the urban spatial structure, this becomes "life frame". This way the spatial-perceptive layer is formed.

The urban organism as system is in continuous transformation is dynamic. This dynamic has been identified quantitatively with map analysis by Petrişor and by Armaş et al. The latter looks to another kind of rural landscape, the one of the forests, with a different productive function than agriculture. However, the presence of urban forests can be occasion of events in cities, as, for example, the performance of Xenakis's Persepolis in the castle garden in Karlsruhe, Germany, or the restoration of the forest-like castle garden in Carei, Romania, shows. It is an element of green infrastructure. If Petrişor only uses GIS for this analysis, Armaş et al. employ GIS for superposing the results from historical map analysis. Analysis which we see at Monteiro for photography or at Meder and Kripper for historical archive research is present. Satellite imagery, which is a way of technical, objective photography, is included in the analysis. The temporal development of the landscape proves vulnerable to human activities which shape it as cultural landscape.

The code of dynamic transformation is what Popa is looking for in the second paper, the dynamics resulting like in the interdependence between urban form and urban structure. She points to the issue of speculation, wanting to win construction land from agricultural land, but we saw in the other two mentioned papers that the forest is endangered by wanting to win agricultural land as well. Through winning construction land from the agricultural land through urban extensive development, the landscape of the periphery appears, from which Stan talks. As such, the periphery appears as a result of the dynamic dialogue between the urban and the surrounding cultural landscape of agriculture or rural development. Although we do not have fortifications anymore to stop uncontrolled extension at the periphery, a green belt could be used as a new kind of fortification Koenigs (1991) recognised. It is the limitation of this extensive development which Constantinescu and Hărmănescu aim for when proposing the improvement in the urban quality of life in cities. Improved quality of the public space would lead to return to the city centre and intensive development with the "fortifications" created in an imaginary way. On this idea of improved public space quality builds also the proposal in Bostenaru of making green walls out of blank side facades of buildings which look to empty parcels, creating a kind of pocket parks—vertical parks in this case. The landscape, which is left outside the city, in the rural area comes back to the crowded centre and its sealed facades. Indeed, also Constantinescu and Hărmănescu see the potential of new technologies in creating public spaces both inside and outside the building and complete with examples from other cities the review performed by Bostenaru regarding green walls. Public space is seen in the examples they give in conjunction with nature, as a "reconciliation ecology" as Francis and Lorimer (2011) called. Francis and Lorimer proposed, as we have seen reviewed in Bostenaru, a participative scheme for this kind of public space. Constantinescu and Hărmănescu touch the idea of the network of public spaces necessary in the city and link this to memory. Indeed, if we would follow Lynch's (1960) approach, a mind map leads to the appropriation of the "heritage habitat" (Gociman 2006) and its persistence within transformations, catastrophic or not. The mind map can consist of a network of buildings as proposed by Bostenaru and Dill for the Modernist boulevard, or by the related approach from the Bucharest Horticulture faculty for Alternative Magheru, which aims at connecting this linear boulevard to the public places in the vicinity. Bostenaru did a project of linking pocket parks this way for Heidelberg in a project presented in Bostenaru (2010b) under the guidance of the landscape architect Henri Bava presented in Bostenaru and Dill. A network trace to make textures cooperate this way is what Constantinescu and Hărmănescu are pleading for. Constantinescu and Hărmănescu identify public spaces as living spaces and as such the space container for activities which build the contained in the morphologic decomposition of the urban system. The zone itself can build a public space in the integrity of its texture. As we have seen, different places of the first type proposed can be born here through the superposition of, as in the example given, a pocket park network. On the other hand, in seeing urban landscape as cultural landscape, Constantinescu and Hărmănescu negate the figure/ground duality from traditional urban planning, both as the one seen by Sitte (1889) and on which traces concept of Bostenaru and Dill builds and as the one see by Le Corbusier when inversing it. A meshwork of interactions appears instead, as highlighted through materialisation of the mind map in the example given. All the pedestrian ways which can be traced as seen in Marina et al. (2011) create new "streets" based on contents, and these are different from a materialised map in pavement textures as in the project we presented above, or in the traces on a piece of paper or in an urban game as proposed by Bostenaru and Dill, but are best mapped with digital technology as by http://www.comob.org.uk/. Unlike public participation GIS which is crowd-sourced mapping, COMOB proposes the mapping of urban traces via GPS in order to see how landscape is being investigated as "ipod-patch". This can be only limitedly implemented in the aftermath by adding GPS data as Bostenaru and Dill do, but are instead calling for revisiting the sites and prove that the city is a labyrinth in which we do not have an Andromede file to guide us in a certain succession. Everyday, itinerary is different. Each perception is temporal. The mind map of every individual is different, as are our memories. But superposition can lead to the creation of landmarks as opposed to zones and boundaries at Lynch (1960), in the way in which Internet network searches lead to some more and some less highlighted words.

If Popa approached photography using philosophy as research method in landscape architecture, Brandt goes further in the employment of philosophy. She publishes an interview with a philosopher on public space, emphasising not so much participation, as the fragility resulting from this. These approaches resulted into creation, as design or as exercise in a series of workshops. Such a cultural landscape is the expression of urban culture, as Stan describes. Stan identifies thus relationships in the urban landscape, between architecture and landscape architecture, which are followed by more papers. Returning to the scale of Petrişor's research, and dealing with a landscape related to Popa, Hărmănescu rediscovers the relationship between the landscape created by the built substance and that created by nature—in other words, the cultural landscape. Actually, Popa in her second article sees the agricultural landscape being a rural landscape as well. She also approaches the idea of memory, like Popa, but the memory in Hărmănescu's approach is immaterially conserved, as traditions in building the dwelling or the public space, which is a question of identity. Hărmănescu proposes the cultural route as means for displaying the cultural identity, a subcategory of the cultural landscape, to tourists. It is what the paper of Bostenaru and Dill deals with, in an extended way. The cultural route identifies elements which can be read in both ways, like at Popa or like at Bostenaru and Dill, depending on the rather built or rather nature character of the landscape displayed at the nodes. She even names the landscape "a living map", combination of lines and dots, the later being built by the spaces, which will be subject of our further research in applying network analysis to landscape. Enache and Căplescu also looks at the cultural landscape, identifying the expression of the human and the nature in the urban landscape. Media are mainly expression of the human. Monteiro looks at cultural identity, as Hărmănescu does, but in a nationalist context, as for Meder and Krippner. In fact, this cultural nationalism question is not far from the link between tradition and innovation (Bostenaru 2010a) and Monteiro makes indeed reference to Hobsbawn (Hobsbawn 1994). Along with the observations of Popa, photography is "a system for dividing, measuring, describing, and analysing the physical world". Through these attributes, it becomes an instrument of political power. For Monteiro, different landscapes are perceived through the lens by photographers of different nationalities, the minority and the ruler. Photography, a new medium in the nineteenth century, has a presumed objectivity compared to the drawing. But in reality, the feelings to be evoked by the landscape depicted reflect a "picturesque" version of the cultural manifestation. Bostenaru also proposed creative design, like Brandt, in order to explore the contribution of landscape, namely increasing the green surface in the city by vertical gardens, not only on the façade, but also in the interior space. Through this, microlandscapes are created, which stayed seldom in connection with landscape urbanism, but which have been investigated by Majuru (2008) in photography, thus returning to the initial method of investigation this time through the scale. If Popa applies Lynch analysis in perceiving the landscape, Meder and Krippner mention the way Viennese architects and garden architects employed the approach by Camillo Sitte in their design with nature. The link can be seen in Bostenaru and Dill, where the way urban space perception is represented is linked between Lynch and Sitte and transposed in the digital interpretation of layers, different from GIS, for landmarks. Urban traces, as proposed by Bostenaru and Dill or by Hărmănescu, are for pedestrians (Marina et al. 2011). The paper by Negulescu is dealing with how to gain back the space from vehicles to pedestrians. Because the image of the city is different, if contemplated as pedestrian, at a Middle Ages scale, or from the vehicle, from the speed. Both Negulescu and Bostenaru relate to the landscape of the street, which, as we will see, is the best expression of the urban form if we see the city as urban organism.

Apart from photography, common to many papers is also employing philosophy as research method in architecture. Many papers include the analysis by Lynch (1960) (Popa, Hărmănescu, Enache and Căplescu, Bostenaru and Dill) in their methodology approach for planning. As such, strategic planning different from traditional regulations is promoted. Bostenaru and Dill and Negulescu even make an incursion in the background of the origins of the user and participation-oriented strategic planning which rose from the critique of the functionalist cities constructed following the Charter of Athens (1933).

We will deal with the background for some of these aspects in detail.

3 Highlighted Selected Conservation and Intervention Approaches

3.1 Real Conservation and Intervention

Petrişor approaches the need for conservation of natural heritage outgoing from quantitative GIS analyses and visualisation of land cover change.

An approach to intervention is proposed by Brandt, who made design exercises to emphasise the participation component of the public space. Also, Stan, in her paper on urban peripheries, proposes intervention. Since the terms "diagnosis and treatment" in conservation derive from medicine, so is the approach she proposes. Intervention is also proposed by Negulescu, to transform the vehicle circulation arteries to pedestrian circulation and thus return to the landscape as perceived by Lynch (1960) and Sitte (1889), which we described in the previous section.

Constantinescu and Hărmănescu proposed intervention as well. Like Brandt, they deal with public space, and like Negulescu, Bostenaru and Enache and Căplescu, the subject of their research is the urban texture. By taking urban texture as subject of their investigation, they relate to the cybernetic analysis of the city after Sandu (1988). Urban textures "sub-systems of the life-frame elements able to respond to the functionally requested situation, an analytic instrument of morphological decomposition of the urban structure in urban theory" (Sandu 1988). On the other hand, the approach to the texture leads us to the issue of the street approached by Bostenaru and by Negulescu. We have seen that the street is the affirmation of the urban character. The "character-of-a-zone" is defined by all the texture categories encompassed and by the texture with the highest land use value. This dominant character is given by the prior function in the zone, which co-operates with others through the unifying homogeneity of the structure, assuring the heterogeneity of the "urban life" and "urban frame". And approach to the street also links to studies of vulnerability since Balamir (2004) in his mitigation approach urban master plan of Istanbul tells that the scope of "risks in urban texture" includes the determination of risks in the "differential formation of urban fabric". And the approach to the street can be seen in a vulnerability analysis as Bostenaru (2005) shows. The intervention is dictated by necessity in order to reduce vulnerability. Alexander et al. (1977) remarked that there is a size of a place in order to be used by the crowd as Constantinescu and Hărmănescu desire. The conditions to achieve this form a network itself. In their conclusions, related to this approach by Alexander et al. (1977) on the relationship between the morphology of the public space as container and the activity taking place in it as contents, Constantinescu and Hărmănescu call for participation as well. It is the user who dictates how public space is shaped, just like in the interview of Brandt. The research methods were different (case study analysis versus philosophy and interview), but the conclusion is similar. And Alexander's et al. (1977) approach has been one of the first participative approaches in history, creating a manual to enable people to design towns, buildings and constructive systems themselves in order to be pleasant to inhabit.

Popa proposed a method to identify how moments in the evolution of the landscape can be preserved in photography. Hărmănescu instead proposes revitalisation of a landscape of a related type (rural landscape/agricultural landscape). It is an intervention that aims at the revival of a former state, the state which might be conserved in photography but is surely conserved by other memory means that shall be restored. Apart from that, Hărmănescu proposes participation in cultural tourism and as such relates to the mentioned public space intervention. Tara Haţegului should become a living landscape. Extending the approach of Petrişor from geosciences and working at the same scale, Hărmănescu proposes to apply the concept of GeoPark to the case study she is dealing with. Preliminary discussions presented to us the GeoPark of south-west Germany (Vogt and Megerle 2006). The dialogue between conservation and intervention becomes that between traditional and modernity, a dialogue which can be well decomposed by means of the philosophical approach of "deconstruction" (Bostenaru 2010a). They do not mean to be antagonic, but treat in dialogue. This is an idea which builds the background also of Monteiro's work, as we have shown.

3.2 Conservation in and of Media and Architecture

Several papers deal with the relationship between the architecture object and the landscape (Enache and Căplescu, Bostenaru, Meder and Krippner). As such, for Enache and Căplescu, the architecture object is support for the "media landscape". For Bostenaru, it is support of the so-called vertical garden. The vertical garden can consist under some instances of the drawn landscape or even the media landscape, as the paper shows. Through the drawn landscape of Art Nouveau, this landscape on the walls of buildings relates to what we have missed when only analysing perception through photography and not through drawing. Through the projected landscapes in media installations, it relates to the large-scale media landscape. Through the fact that plants are represented on buildings, it relates to today's vertical garden, as explained by Bostenaru. Returning to the landscape of plants, it relates to the climbing vegetation in which Modernist architecture has been seen by Meder and Krippner. The paper by Bostenaru makes direct reference to this, by exemplifying three buildings from the interwar time which employed climbing plants or trees in Hundertwasser manner and pointing a way to review literature of German language which makes reference to green walls from the nineteenth century to present. However, the approach is not so consistent as in Art Nouveau with nature. Maybe for this reason, this time the Reseau Art Nouveau Network (http://www.artnouveau-net.eu/) is running a project called Art Nouveau and Ecology with a current exhibition (2013, Brussels Info Centre) with the representation of nature in Art Nouveau. But the exhibition does not highlight how vegetation is being represented in Art Nouveau. Instead, the focus lays on nature represented through its elements: water, fire, earth, air, as in Oriental mythology. So the drawing of nature can be seen in this context as imaging of landscape, such as the one done with photography on which we will detail, but this image is projected on buildings or stained windows. In drawing the image of landscape or only that of plants, the artists concentrate on the essential lines of this, just as Popa identified in macroscale using Lynch (1960) concepts. Iconic representations are born reminding of the sacred painting of the Middle Age, and this is not only through the partial medium which is the stained glass derived from churches. The building becomes canvas for landscape representation. Media landscape as described by Enache and Căplescu can be seen as a use of buildings as canvas for different messages, one of the ways of achieving this being the projection of images as in the ZKM installation or other installations proposed by Bostenaru.

3.3 Conservation by Means of Photography

As Popa identifies, photography has to do with memory, discovering the essence of a landscape at a specific moment in time. As such, it is the conservation of that moment, a virtual conservation of the landscape. Bostenaru (2011a) was a paper about the conservation of catastrophe memories by means of photography, since those are ruins of the moment, created in a moment and eradicated through reconstruction efforts. We will deal with this in our future book. Bostenaru (2011b), instead, dealt with photography of the street in a philosophic interpretation, a means employed by many papers of this book. The street is a landscape and expression of the urban form. Sandu (1988) affirms that "out of all urban spaces, the street is available as the most adequate location for the affirmation of the urban character". Popa looks to what within the photograph and how the tension lines within the landscape are found and interpreted and not to the 2D representation as such in photography, as Rene Berger (1963) makes when analysing paintings. We tried in a time to apply Rene Berger's (1963) method for interpreting a landscape expressed in a poetry piece by the Romanian poet Bacovia: the lines of the trees in winter etc. Photography is seen as description medium. Instead, Eftenie (2010) saw photography as tool for design as well, in architecture and urban planning, and is thus also related to the approach by Popa.

4 Outlook to Further Studies on Sustainability and Resilience

Apart from the study of the landscape by means of photography, the study by means of drawing shall be investigated in the future, since this highlights exactly those considered to be constitutive elements of landscape by Popa. In fact, in one of the quotes, it makes reference to painting. However, we intend to deepen this study of the role of photography in a future volume at the publisher on catastrophe photography. There we deal with historic photography like Monteiro but also perform historical network analysis in order to discover the correlations between site and photography method in historic photography. The method of Popa can be put in dialogue with landscape character studies, including those involving participation and "citizens as sensors" as shown by Bostenaru and Dill or more spread methods like PhotoVoice.

In further studies, we will also deal with visualisation, of how characteristics of the space can be expressed, for example in datascapes, in virtual reality. To this serves the Network for Digital Methods in Arts and Humanities in which we formed a cluster of Digital Landscape Architecture, aligned with the international initiatives in the field. It includes both creation of landscape, as approached by this book, and also analysis of landscape, both technically and by philosophic/artistic means for planning purposes. Editing of a book on that subject, more precisely called "Space and Time Visualisation" has already started.

From our work at the Canadian Centre for Architecture, we learnt to deal with "datascapes" as well, not just with photography. A fellow researcher (Chapel 2013) was presenting this. In our future book, we will extend the concepts from Bostenaru and Dill to datascapes from GIS and Google Earth using DataAppeal, software inspired by MVRDV datascapes (Amoroso 2010). Mihăilă and Enache (2012) interpreted the same interdependence between densities expressed in data and landscape, as preliminary work for a project application.

We focused in this book on the ecological–aesthetic–social function of landscape. But in order to protect landscape, which is endangered by speculative developments in real estate, the participative approach shall be enhanced, "ideas shall be formed in the heads of people", since this protects better than legislation (Koenigs 1991). And also the economic function shall not be neglected. If we look to analyses by Petrişor and Armaş et al. directly, and those by Hărmănescu indirectly, the forest is an endangered item. Just in 2012, the League of Romanian Students abroad organised a colloquium on protecting forests in order to protect from natural hazards such as floods to which we participated. Popa mentioned the effect of natural disasters on the dynamics of the rural landscape, an issue which deserves deepened study in further volumes, in either the envisaged or dedicated series, since also Crăciun in her strategy description (Crăciun 2014) and Stan identify the vulnerability characteristics of landscape. Armaş et al. study can build a basis for a multicriteria analysis on vulnerability of the landscape already in this volume, from the point of view of geography. Between analysis and strategy, looking at an approach to disaster management divided into hazard, vulnerability, risk assessment and risk management, the vulnerability of landscape deserves further study. For this, J. Barkley Rosser Jr. developed a concept on treating forests as dynamically complex ecological–economic systems (2012) presented in frame of the COST Action IS1104 at the NED conference in Sienna. We try in the same framework to look at the economics of green spaces, as also touched in the article as research question, since green spaces enhance the value of adjacent space and should be protected from speculation approaches. It is a conservation–intervention balance per se, as subject of the book.

References

Alexander Ch, Ishikawa S, Silverstein M (1977) A pattern language: towns, buildings, construction. Oxford University Press, New York

Amoroso N (2010) The exposed city: mapping the urban invisibles. Routledge, London

Berger R (1963) Discovery of painting. Viking Press, New York

Balamir M (2004) Urban seismic risk management: the earthquake master plan of Istanbul (EMPI). In: Proceedings of the 13th world conference on earthquake engineering, Vancouver, BC, Canada, Paper No. 9005

Bostenaru Dan M (2005) Multidisciplinary co-operation in building design according to urbanistic zoning and seismic microzonation. Nat Hazards Earth Syst Sci 5:397–411

Bostenaru Dan M (2010a) Vernacular and modernist housing in Germany and Romania: an analysis of vulnerability to earthquakes. Cuvillier Press, Göttingen

Bostenaru Dan M (2010b) Spaţiul verde redescoperit—Der wiederentdeckte Grünraum. Cuvillier Press, Göttingen

Bostenaru Dan M (2011a) The use of ontology for digital conservation of architecture works after catastrophes. J Appl Eng Sci 14(2):11–18

Bostenaru Dan M (2011b) The street in the photography from the 19th and early 20th century as mean of art or science. e-conserv Mag 19:57–69 http://e-conservationline.com/content/view/999

Crăciun C (2014) The vulnerability of fragile landscape and complex hazards. areas with potential risks and specific types of hazards and vulnerabilities of the natural, anthropic and cultural landscape in Braila County, Romania. In: Bostenaru Dan M, Armaş I, Goretti A (in print) Earthquake hazard impact and urban planning, Springer Netherlands, Dordrecht

Chapel E (2013) Nouveaux médias et persistance du livre MVRDV, Metacity/Datatown, 1999, In: Chabard P, Kourniati M (eds) Raisons d'écrire, livres d'architectes, 1945–1999. Éditions de La Villette, Paris, p 179–204

Deming EM, Swaffield S (2011) Landscape architectural research: inquiry, strategy, design. Wiley, New Jersey

Doucet I, Janssens N (2011) Transdisciplinary knowledge production in architecture and urbanism, urban and landscape perspectives. Springer Netherlands, Dordrecht

Eftenie V (2010) Vocaţia publică a spaţiului arhitectural-urbanistic. Fotografia ca mijloc de investigare în planul imaginii urbane, instrument de operare în procesul de concepţie—proiectare, PhD thesis, Ion Mincu University of Architecture and Urbanism

Francis RA, Lorimer J (2011) Urban reconciliation ecology: the potential of living roofs and walls. J Environ Manage 92(6):1429–1437
Gociman CO (2006) Managementul reducerii riscului la dezastre: strategii de arhitectura˘ şi urbanism. Editura Universitara˘ "Ion Mincu", Bucharest
Hobsbawn E (1994) The nation as invented tradition. In: Hutchinson J, Smith A (eds) Nationalism. Oxford University Press, Oxford, pp 76–82
Koenigs T (ed) (1991) Vision offener Grunräume: Grüngürtel Frankfurt = Vision on Open Green Spaces: GreenBelt Frankfurt (in German). Campus-Verl, Frankfurt/Main
Lynch K (1960) The image of the city. MIT Press, Cambridge
Majuru A (2008) Micropeisaj, public talk. Reviewed at http://www.natkat.org/project_results/event2.html
Marina O, Ivanovski J, Karanakov B, Ivanovska A, Topalovska I, Stavric M (2011) Modelling pedestrian mobility as designer's tool in urban protected areas. In: Proceedings of the 11th international conference "reliability and statistics in transportation and communication" (RelStat'11), Riga, Latvia, pp 408–418. 19–22 Oct 2011
Mihăilă M, Enache C (2012) Density and landscape_working paper. In: Mitrache A (ed) Cultural landscape, architecture trends, 120 years of higher education in architecture, Editura Universitară Ion Mincu, Bucharest
Rosser B (2012) Special problems of forests as dynamically complex ecological-economic systems. In: Proceedings of the international conference on new frontiers of forest economics ETH, Zurich, Switzerland, June 26–30, 2012. Available online at http://neffe.ch/fileadmin/pdf/neffe/papers/paper.rosser.pdf accessed 28.09.2013
Sandu A (1988) Urban composition, course notes. Ion Mincu University of Architecture and Urbanism, Bucharest
Sitte C (1889) Der Städtebau nach seinen künstlerischen Grundsätzen. New edition: Birkhäuser, Basel 2002
Vogt J, Megerle A (eds) (2006) *Geopark- und Geotourismusforschung—Ergebnisse des Internationalen Workshops in Karlsruhe 2006*. Regionalwissenschaftliche Forschungen/Regional Science 31

About the Authors

Iuliana Armaş is full-time Professor in Physical Geography and Natural Hazards and head of the Center for Risk Studies, Spatial Modeling, Terrestrial and Coastal System Dynamics at University of Bucharest, Faculty of Geography, Romania.

Maria Bostenaru Dan is a researcher, PhD, at the "Ion Mincu" University of Architecture and Urbanism and graduate in architecture, specialization urban planning of the University of Karlsruhe, Germany as well as member of the board of the Marie Curie Fellows Association.

Stephanie Brandt is head at Stephanie Brandt Studio and founder of SPACEPILOTS and is a graduate of The Bartlett School of Architecture, UCL London, UK in architecture.

Cristina Constantinescu assistant professor at "Ion Mincu" University of Architecture and Urbanism, degree in urban planning (2005) and architecture (2010), PhD in urban planning and territorial improvement (2012).

Cerasella Crăciun PhD—Associate Professor Architect, Vice-Dean—Urban Planning Faculty, "Planning and Landscape Design" and "Landscape and Territory" Masteral Program Coordinator, Urban Planning Faculty; Member of the Romanian Register of Urban Planners, Chamber of the Romanian Architects, Romanian Union of Architects, Vice-President of Romanian Urban Planners Professionals Association (APUR), Founding member of CC—Corpus Colosum Group of Transdisciplinarity.

Alex Dill, architecture engineer, is Akademischer Oberrat at the Karlsruhe Institute for Technology, head of the "Interiors and Furniture Design" teaching subdivision of the Institute of Architectural Design, Art and Theory at the Faculty of architecture and also member in the board of DOCOMOMO Germany.

Cristina Iuliana Enache, Associate Professor, PhD architect, at Urban and Landscape Design Department, "Ion Mincu" University of Architecture and Urbanism, Bucharest, graduated as architect, UAUIM Bucharest. She holds a Doctor's degree in Urban Planning and she is the managing partner at ARTTEK CB architecture office.

C. Crăciun and M. Bostenaru Dan (eds.), *Planning and Designing Sustainable and Resilient Landscapes*, Springer Geography, DOI: 10.1007/978-94-017-8536-5,

Mihaela Hărmănescu, PhD, is the present academic staff of Faculty of Urban Planning, UAUIM Bucharest, vice-president on cultural landscape issues of RPER-RO Rencontres du Patrimoine Europe Roumanie, member of Strategic Research Initiative Board Directors, member of Romanian Chamber of Architects (OAR), graduated as architect, UAUIM Bucharest with M. Arch. (Sustainable development—an integrated concept of built space and energetic audit, 2006) and PhD in Urban Planning (Potential and value of the rural landscape as part of the cultural landscape, 2012) and founder of Mihaela Harmanescu architecture office (MH. Bia).

Ulrike Krippner, DI, is a researcher at the Institute of Landscape Architecture (ILA) at the University of Natural Resources and Life Sciences (BOKU), Vienna, Austria, where she graduated in landscape architecture.

Iris Meder is PhD, a free lance curator and researcher at the Institute of Landscape Architecture (ILA) at the University of Natural Resources and Life Sciences (BOKU) in Vienna, Austria, and a graduate of the University of Stuttgart, Germany, in Art History and Literature.

Stephen Monteiro is a professor of visual culture and media studies at The American University of Paris. He is a graduate of the Sorbonne.

Mihaela Hermina Negulescu is PhD (with the thesis "Mobility and Urban Form"), Architect, MSc Urban Planner, Associate Professor at the Faculty of Urban Planning and Coordinator of its "Urban Mobility" Masteral Program, at the "Ion Mincu" University of Architecture and Urbanism, Bucharest.

Gabriela Osaci-Costache (BSc single Hons in Geography, University of Bucharest; MSc Degree in "Environmental Morphohydrography an Land Planning"; PhD Degree in Geography, University of Bucharest) is a lecturer at the Department of Geomorphology—Pedology—Geomatics, Faculty of Geography, University of Bucharest, Romania.

Alexandru-Ionuţ Petrişor, PhD (Ecology), PhD (Geography); specialized in urban ecology, bio- and geostatistics. Currently, Assistant Professor of Ecology and Statistics, "Ion Mincu" University of Architecture and Urbanism, and Senior Researcher and Scientific Director, National Research Institute "URBAN-INCERC."

Angelica Stan is PhD (in urban planning in 2006 at the "Ion Mincu" University of Architecture and Urbanism with a thesis about the landscape of urban peripheries), Architect, MSc Urban Planner, Associate Professor at the "Ion Mincu" University of Architecture and Urbanism, Bucharest, Faculty of Urban Planning, Urban and Landscape Design Department.

About the Editors

Assoc. Prof. Dr. Arch. Cerasella Crăciun studied at the Institute of Architecture "Ion Mincu" in Bucharest, graduating after 6 years of courses at the University of Architecture with a landscape dissertation on the topic "Symbiotic Architecture and the Japanese Gardens." This was the first diploma in the history of the Architecture Institute that explored and integrated architecture, urbanism and landscape in a transdisciplinary way, by transgressing the macro-, mezzo-, and detail levels and including such issues as the intuitive landscape analysis of the context at different urban levels, operations of urban planning, rules and regulations, until the detailed study of the architecture object and of interior design. After finalizing her undergraduate studies, she was enrolled in the postgraduate study program in urbanism, specializing in "Urban Form—Architecture, City, Territory," with the dissertation thesis: "Experiment, Transformation, and Metamorphosis in the Urban Metabolism." At the same time, she continued her

C. Crăciun and M. Bostenaru Dan (eds.), *Planning and Designing Sustainable and Resilient Landscapes*, Springer Geography, DOI: 10.1007/978-94-017-8536-5,

research activity at the National Institute of Research and Development for Urbanism and Territorial Management—URBANPROIECT, as well as her didactic activity as guest and then tenured professor at the Faculty of Urbanism, a new specialization created within the University. During this period, alongside Professor PhD Arch. Angela Filipeanu, she consolidated the making of the first "germ-seed" in the landscape design field in Romania, through her didactic and educational activity, as well as through her efforts of conducting workshops, courses, and landscape design and research activities. All of these efforts were capitalized on in the year 2000 when the decision to create a new department of "Landscape Architecture" (currently known as "Planning and Landscape Design") with an undergraduate program of 5.5 years and integrated master's degree was taken, and the file was turned in at the Ministry of National Education; this department was to be coordinated by PhD Architect Cerasella Crăciun. This proposal was also realized in agreement with and supporting the European Landscape Convention (ratified in Romania through Law no. 451 from 8 July 2002) for the training of new specialists in the understanding of interventions regarding landscape politics, protection, management, and landscape design. Therefore, in view of the Bologna process, Cerasella Crăciun initiated, advanced, and put forward the file regarding the accreditation for the Master's Degree Program "Landscape and Territory" (2 years/120 ECTS credits), a program that she has been coordinating since the year in which it was created (2008).

In parallel to her continuous activity of understanding the landscape domain, not only in the domain of her specialization, but also within the curricula of all the faculties and specializations within "Ion Mincu" University of Architecture and Urbanism, Cerasella Crăciun developed a remarkable teaching activity, with notable contributions and innovative elements, which inspired numerous students, who have become PhD students and assistants.

Graduating the pedagogic course at the University Politehnica in Bucharest, in the department for preparing teaching staff, with an average grade of 10.00 (out of 10.00) led to her thorough study of educational psychology, pedagogy, methodology, human communication, educational sociology, pedagogical practice, and didactic projects. All of this guided her toward the need to materialize and apply new techniques of transdisciplinary education, within the numerous courses sustained ("Landscape Design," "Projection and Restoration of the Natural Heritage," "The Urban and Individual Garden," "Research, Set Design, and Arts within the Landscape"), workshops projects, as well as within some landscape modules, which are addressed to all the students of the university that attend the undergraduate and master's degree program.

Cerasella Crăciun has a PhD in Urbanism, she is Associate Professor, Vice-Dean of the Faculty of Urban Planning (2013), Head of Urban and Landscape Design Department (2011–2013), coordinator of "Landscape Design and Planning" Bachelor's Degree Program and "Landscape and Territory" Master Program at the Faculty of Urban Planning, Bucharest "Ion Mincu" University of Architecture and Urbanism. She is also the one who initiated the process of including The Faculty of Urban Planning in the European Council of Landscape Architecture Schools (ECLAS). Through her creativity and flexibility, from landscape to urban studies, from architecture, to interior design and set design, to geometry and sacred geography research, the author is a transdisciplinary person with Renaissance-type holistic concerns, being involved in the design process, in education and in the research of landscape planning and sustainable urban development fields, but also in transdisciplinary border studies.

Through her doctoral thesis, she elaborated on the analogy between the biological organism and Landscape seen as a living organism ("body" areas—city organism), through searching for new methodologies, theories, and methods of approach, diagnosis and analysis in urban architecture and landscape, considered unconventional science as opposed to current approaches in landscape, urban, or architectural fields. In this sense, the landscape and the urban metabolism are seen as a process in direct relationship with Man, the Planet and the Universe through a Holistic approach. The theories and the unconventional methods come from: the science of complexity and the morphogenetic field, fractal and construction theory, computer science, music and the science of sounds, sacred geometry, and geography.

Member of The Romanian Register of Urban Planners, The Romanian Order of Architects, The Professional Association of Urban planners in Romania, founding member of the transdisciplinary group CC Corpus Callosum, and "Scarlat Demetrescu" Society, member of the International Federation for Housing and Planning. She is also a DOCIS expert—for the development of an operational qualification system within the higher education system in Romania, member of Implexus—Education, Critique, and Complexity, as well as a member of numerous editorial colleges, landscape scientific committees, and specialized juries for national contests. Her fields of interest are Landscape Planning, Cultural Landscape, Transdisciplinarity, and the Sciences of Complexity, the Urban Metabolism, Territorial Planning, and Public Politics in Landscape, Sacred Geometry and Geography, Lithopuncture, Unconventional Arts in Landscape, the Scenery Design of the Public and Built Space, the Philosophy and Perception of Landscape, Archetypal Science.

Cerasella Crăciun obtained numerous awards, nominations and distinctions, both personally and together with the team she conducts, such as: The Award in the Field of Sustainability for "The Educational Initiative in the Field of the Year's Green Buildings" given to the Faculty of Urban Planning (team award) by the Romanian Green Building Council (2013); the Ministry of Regional Development and Tourism offered a team award within the National Architecture Biennale with a "Diploma for the contribution to the consolidation and development of strategic planning of urban areas and territory" awarded for the Section of "Public space, quality of urban life, and Landscape as a fragile resource—The mezzo-landscape and the natural, anthropogenic, and cultural landscape in Bucharest" in the framework of the "Integrated Urban Strategy for development of the city of Bucharest and of its support and influential territory—Strategic Concept Bucharest 2035" (2012); "Honor of Understanding Performance and Dedication" for "Best Gardens Designer Exhibition—'Le Notre' Jardins a la Francaise," Chateau de la Huardiere, France (2010/2011); winner of the Cultural Project "Lost Gardens," after an organized selection of the Union of Architects of Romania (2010); the prize for Landscape Management of the Romanian Register of Urban Planners for the "Protection of the Bucharest Parks—Herastrau, Cismigiu and Carol (2008); selection by an international jury for the Publishing Section for the book "Urban Metabolism, An Unconventional Approach to the Urban Organism," within the Architecture Biennale (2008); winner of "Urban-Landscape Design and Architecture Management—Red Islands—Danube's Delta" Contest, UAUIM (2000); winner of "Bucharest to the North of the Lakes" contest, UAUIM (1995).

She is the author of four books (and three more which are being written at the moment), of over six chapters in specialty books (other four are being written), of over 50 articles, of over 50 conferences and workshops. She is also the author of nine exhibitions, co-author of four exhibitions, and curator and coordinator of over 20 exhibitions. She coordinated over 30 workshops, seven research grants, over 60 studies in landscape projects, over 30 studies and projects in urbanism, territorial management, and urban planning. We can add to all this a set of architectural projects, projects of object and interior design, set design, jewel collections and fashion design, graphics, watercolor, and painting exhibitions.

In parallel to her didactic and research activity, Cerasella Crăciun developed her activity in the field of the legislation governing the landscape field, through the promotion of the Landscape European Convention, through comments and additions made to the Law of Green Spaces in Romania, and through her public critical attitude toward the destruction of green spaces, of the landscape heritage, and of the cultural, immaterial,and rural landscape. She also has a long lasting activity in the field of landscape design and of regulatory documents for landscapes through regulations, methodologies, and landscape planning projects. Thus, she put together the only morphological, structural, and configurative study at a national level that researches the landscape in all its components (natural, anthropogenic, and cultural), at all its territorial and urban levels, from macro- to mezzo- and detail level landscape, as well as through all its typologies.

Dr. Dipl.-Ing. Maria Bostenaru Dan obtained a degree in architecturee, specialization in urban planning, from the Universität Karlsruhe, Germany in 1999. She holds certificates in additional studies in "Working with multimedia" and "Project Management" from the same university. In 2012, she obtained her doctorate with the topic "Interwar architecture with reinforced concrete structure exposed to seismic hazard. Intervention in Romanian and Italian context" at the "Ion Mincu" University of Architecture and Urbanism, Bucharest, Romania. With the Collaborative Research Center "Preservation of historically relevant constructions" she did a building survey in Poland in 1998 and was employed as research assistant at the CRC "Strong earthquakes" in 2000–2001, both at the Universität Karlsruhe. With a three year DFG (German Research Foundation) scholarship in the Research Training Group "Natural Disasters" in Karlsruhe and a 6 months stay at a Marie Curie Training Site in Pavia, Italy, research was done on "Applicability and economic efficiency of seismic retrofit measures on existing buildings" 2000–2004. She was experienced researcher on the project "Preservation of historic reinforced concrete housing buildings across Europe" (CA'REDIVIVUS), a 2 year Marie Curie Intra-European Fellowship in Pavia 2005–2007. She returned to Romania with a Marie Curie Reintegration Grant on "The innovation in the plan of the current floor: Zoning in blocks of flats for the middle class in the first half of the twentieth century" (PIANO) 2007–2010, which was selected by the European Commission as promising result under the headline "Building the future on lessons of the past." After working a few months as research associate, she is employed since 2008 as scientific researcher at the "Ion Mincu" University of Architecture and Urbanism, permanent position, first at the Department for History and Theory of Architecture and Heritage Conservation, since 2011 at the Department for Urban and Landscape Design. There she was involved in CNCSIS funded research on "Arts, Urban Communities, Mobilization—the social insertion of the artistic and architectural project" and in a Union of Romanian architects co-funded project: the Tzigara-Samurcas archive at the first department and then has been involved in the "Academic Network for Disaster Resilience to Optimize educational Development" (ERASMUS LLP), the Le Notre ERASMUS LLP on landscape planning as deputy delegate for the university as associate member and in the CNCS project "Urban blocks in central protected area in multiple hazard approach assessment: mapping and strategies for risk mitigation. Case study:

Bucharest—Destructured zone by razing occurring in communist period." During this time, she was grantee in an FP6 support action: the "Cultural heritage research meets practice" conference and of an EEA Grant on advocacy planning, apart of other travel grants from conference organizers. She kept doing consulting on earthquake engineering in the CNMP project lead by the University of Bucharest "Multihazard and vulnerability in the seismic context of Bucharest city" 2007–2010 and is employed since 2012 in a CNCS project of the same on "Spatial and temporar patterns of vulnerability" as well as through the work with the "World Housing Encyclopedia" (project member since 2001, editorial board member 2003–2006, participant to the taxonomy project in frame of the Global Earthquake Model initiative) and co-teaching the course "Protection of localities against risks" at the "Ion Mincu" University for Architecture and Urbanism. She is an management committee member at the COST actions "Semantic enrichment of 3-D city models for sustainable urban development" (since 2009), in frame of which she spent 5 weeks at the Landscape Department of the University of Algarve, Faro, Portugal to do research on earthquake impact representation in 3-D models of Lisbon, and "The EU in the new complex geography of economic systems: models, tools, and policy evaluation" (since 2012), in frame of which she spent 5 weeks at ISCTE, Lisbon, as well as of the COST action gender STE (since 2013). She is steering committee member of the ESF Network for Digital Methods in Arts and Humanities (since 2012), in frame of which she acquired funding to organize a crosswork group meeting in Bucharest in November 2012 and spent 2 weeks at the Karlsruhe Institute of Technology, about which she reports in this book. She had Canadian Center of Architecture support grant in 2010 to do research on realism in nineteenth century photography of natural and man-made catastrophes. Recently, she has been awarded a DOMUS scholarship by the Hungarian Academy of Sciences, where she is the member of the external public body to do one month research at the Corvinus University of Budapest, Faculty of Landscape Architecture and a postdoctoral one month grant for the Dumberton Oaks research institute on landscape architecture of the Harvard University, forthcoming both in 2014. Maria Bostenaru Dan has been member of a number of professional associations, also occupying leading positions, starting with student leadership positions. In 2003, she was the member of the Advisory board of the Marie Curie Fellows Association, and since 2011, she is the member of the Administrative board of the same. She is the member of the international scientific committee of ICOMOS ISCARSAH (analysis of structures, since 2010) and pending confirmation for ISCEC (economics of conservation). She has been in the scientific committee for several conferences. She was session organizer numerous times at the European Geosciences Union and its forerunner, and special issue editor for the official journal of the European Association for Earthquake Engineering. During her time as editorial board member at the World Housing Encyclopedia, a summary report book has been released. She has been peer reviewer for a number of journals and conferences and is editorial board member for the International Journal of Biourbanism, of the Romanian journal "Urbanism. Arhitectură. Construct¸ii" and of the Canadian Journal of Engineering Management. She is an expert evaluator for Marie Curie Career Integration

Grants, a funding scheme of the European Commission, for COST transdomain proposals as well as for cultural projects of the Administration of the National Cultural Funds, Romanian Ministry of Culture, in the fields "Architecture and Design" and "National Cultural Heritage." She has been accepted as evaluator for the Marie Curie Intra-European Fellowships, for ERASMUS and for Culture programmes. Maria Bostenaru Dan has over 150 publications and presentations, including ISI journals, seven authored books in German and English, one former edited book at Springer, and edited historic manuscript and has been member of collectively done books such as on urban regeneration, Twenty-first century housing in Romania, historic photography, and research on the cemetery of Bucharest. Her research domains are housing, earthquake engineering, construction management, reinforced concrete, early twentieth century architecture, participatory planning, photography, and digital archives. Most cited are her works on housing and decision systems.

From the ECLAS award nomination written by arch. Cristina Enache, PhD.

Index

C. Crăciun and M. Bostenaru Dan (eds.), *Planning and Designing Sustainable and Resilient Landscapes*, Springer Geography, DOI: 10.1007/978-94-017-8536-5,

MIX
Papier aus verantwortungsvollen Quellen
Paper from responsible sources
FSC® C105338

If you have any concerns about our products,
you can contact us on
ProductSafety@springernature.com

In case Publisher is established outside the EU,
the EU authorized representative is:
Springer Nature Customer Service Center GmbH
Europaplatz 3, 69115 Heidelberg, Germany

Printed by Libri Plureos GmbH
in Hamburg, Germany